SCOPE 15

Environmental Risk Assessment

Executive Committee of SCOPE

President: Professor G. White, Institute of Behavioral Science, University of Colorado, Boulder, Colorado 80309, U.S.A.

Vice-President: Professor S. Krishnaswamy, School of Biological Sciences, Madurai University, Madurai 625021, India.

Vice-President: Professor G.A. Zavarzin, Institute of Microbiology, USSR Academy of Sciences, 117312 Moscow, U.S.S.R.

Secretary-General: Dr. F. Fournier, Inspecteur Général de Recherches, ORSTOM, 24 rue Bayard, 75008 Paris, France.

Treasurer: Dr. A.H. Meyl, Deutsche Forschungsgemeinschaft, Kennedyallee 40, 53 Bonn, West Germany.

Members

Professor M.A. Ayyad, Botany Department, Faculty of Science, University of Alexandria, Moharran Bay, Alexandria, Egypt.

Dr. H. Egan, Laboratory of the Government Chemist, Cornwall House, Stamford Street, London SE1 9NQ, U.K.

Professor V. Landa, Deputy Secretary General, Czechoslovak Academy of Sciences, Národni 3, Prague 1, Czechoslovakia.

Professor J.W.M. La Rivière, International Institute for Hydraulic and Environmental Engineering, Oude Delft 95, P.O. Bx 3015, 2601 DA Delft, The Netherlands.

Dr. T. Rosswall, SCOPE/UNEP International Nitrogen Unit, Royal Swedish Academy of Sciences, Fack S-104 05, Stockholm, Sweden.

Editor-in-Chief

Professor F.J. Fenner, John Curtin School of Medical Research. The Australian National University, P.O. Box 4, Canberra, A.C.T. 2600, Australia.

Theodore Lownik Library
Illinois Benedictine College
Lisle, Illinois 60532

SCOPE 15

Environmental Risk Assessment

Edited by
Anne V. Whyte and Ian Burton,
Institute for Environmental Studies,
University of Toronto, Canada

Published on behalf of the
Scientific Committee on Problems of the Environment (SCOPE)
of the
International Council of Scientific Unions (ICSU)
by

JOHN WILEY & SONS
Chichester · New York · Brisbane · Toronto

3 63.7
E 61

Copyright © 1980 by the
Scientific Committee on Problems of the Environment (SCOPE)

All rights reserved.

No part of this book may be reproduced by any means, nor
transmitted, nor translated into a machine language without the
written permission of the copyright holder.

British Library Cataloguing in Publication Data:

Environmental risk assessment. — (SCOPE; 15).
  1. Environmental health — Evaluation
  I. Whyte, Anne V II. Burton, Ian III. International Council
  of Scientific Unions. *Scientific Committee on Problems of the
  Environment*
  614.7   RA566   79-42903

  ISBN 0 471 27701 0

Typeset by Photo-Graphics, Stockland, Honiton, Devon and
printed in the United States of America.

Contents

vi

Foreword

The International Working Seminar on Environmental Risk Assessment, June 8-14 1977 was organized by the Scientific Committee on Problems of Environment (SCOPE) of the International Council of Scientific Unions, and the Hungarian Academy of Sciences. The Biological Research Institute of the Hungarian Academy of Sciences at Tihany, located on the shore of Lake Balaton, was chosen as a host to the seminar.

Lake Balaton, on the shores of which there are over 500,000 holidaymakers on weekends in summer, is one of the most precious natural resources of Hungary. In addition, intensive agricultural production and food processing have been developed in the lake's surroundings. Consequently, during the past decades eutrophication has speeded up. This urged the respective Hungarian scientific bodies and regional administrative organs to elaborate a long-range programme for the slowing-down of eutrophication through preserving the lake's nutrients and for the solution of other environmental problems (i.e. noise, air pollution, etc.) of the resort areas.

The example of Lake Balaton demonstrates that certain conflicts between nature and society may arise during the process of economic development. Generally, there are no measures which have only favourable impacts, but concrete actions can be taken only after the careful assessment of favourable and unfavourable effects.

Usually, considerable funds are needed for every operative action enhancing the values of the environment. Since there is no country which would have unlimited financial resources for these purposes, the selection of top priorities and their ranking through the estimation of the risks of unfavourable impacts is of the utmost importance. A powerful environmental policy cannot function without risk assessment. This is particularly important in the case of countries where the difficult task of industrial development is being carried out in our time.

Environmental risk assessment is a new method and means for the elaboration of the economic and science-based alternatives of processes applied in protecting human health. This was why the Hungarian Academy of Sciences was pleased to be a host to the above-mentioned seminar, where experts from 23 countries could have a beneficial scientific exchange of views and could part in mutual respect and friendship.

I. LÁNG
Deputy Secretary General
Hungarian Academy of Sciences

List of Figures

List of Tables

Editors' Preface

The Scientific Committee on Problems of the Environment (SCOPE) and its parent body ICSU are concerned primarily with scientific problems. It has been recognized from the outset of SCOPE's activities that *environmental science* cannot be effectively pursued except on an interdisciplinary basis involving different scientific specializations from the physical, biological and social sciences. Also involved are economics, law, and a wide range of other areas of special knowledge such as community medicine, environmental psychology, communications and information theory, to name only a few.

It has been SCOPE's objective to bring appropriate groups of specialists together to pool and develop their collective wisdom on topics of concern to the international scientific community and to governmental organizations concerned with the management of problems of the human environment.

An early action of SCOPE in furtherance of this objective was to establish a project entitled 'The Communication of Environmental Information and Societal Assessment and Response' (Project 7). This project has received sustained support and interest from the United Nations Environment Programme, and has involved the direct participation of nearly 100 scientists from more than 60 countries, as well as many others indirectly.

The idea of Project 7 was to begin to examine the science and society interface in terms of the communication of information from scientists to governments and the general public. It was felt that many examples of miscommunication could be readily observed and SCOPE scientists wanted to see how the situation might be improved. Often scientists were conscious of issuing warnings about environmental problems and seeing them ignored or overlooked. On other occasions scientists felt that tentative and uncertain scientific findings or hypotheses were exaggerated out of all proportion by oversimplification or distortion in the news media, leading sometimes to undue alarm on the part of citizens and overhasty reaction by governments.

At an early 'project shaping' meeting held by invitation at the Holcomb Institute, Butler University, Indianapolis, USA (August 1973), it was decided to focus upon the concept of environmental risk as an approach to the problem of communication. This gave a more specific focus to the subsequent work and avoided the danger of being too pre-occupied with the more melodramatic aspects of 'environmental crisis' reporting. The term 'risk' also served as a useful starting point for interdisciplinary communications, and it quickly became apparent that 'risk assessment' could be seen as a newly emerging transdisciplinary field in need of development.

xv

The first phase of the work of Project 7 was therefore designed to create an overview or state-of-the-art report on risk assessment and to work towards a common language usable by all or most scientists and readily understandable by non-scientists. A meeting was held at the Wood's Hole Oceanographic Institute (April 1975) under the title *Comparative Risk Assessment of Environmental Hazards in an International Context*. The major outcome of this meeting is the SCOPE Report 8: *Risk Assessment of Environmental Hazard* prepared by Professor Robert W. Kates. SCOPE 8 serves as an introductory overview of the rapidly growing field of risk assessment. It is also the main point of departure for the present report.

Human response to environmental risk has been, and to a large extent remains, in a disorganized state. Massive expenditures of public funds are accepted in some areas to make very small marginal improvements in safety or risk reduction, while much larger risks are tolerated, accepted or even welcomed in other areas. There is a high degree of inequity and irrationality in the distribution of risks among different activities; among different types of employment; between different regions of countries, and among nations themselves. Scientific attempts to assess and explain environmental risks are complicated by intruding issues of international trade, by legal and jurisdictional differences, by irrational fears, by sectional and conflicting interests, and by short-sighted policies and actions.

We have hopes that this situation can be dramatically improved if the ideas and methods of risk assessment are sorted out, and to some extent codified and agreed upon. This will not remove conflict, or guarantee that no crises will occur. It should ensure a much larger element of rationality in the system however, and a more professional stance on the part of the managers. We believe that this is what is needed, and what is happening, in environmental risk assessment.

The aim of this report is much more modest. Its purpose is to convey some of the things that need to be done and some of the questions that have to be faced in implementing a risk assessment approach. It is not a how-to-do-it manual or handbook. The diversity of risks in the environment and the uneven quality of experience means that much more evidence must be gathered, sifted and critically examined before such an authoritative handbook is possible. This text moves as strongly in that direction as appeared practical to the groups and individuals at the time as they participated in the preparation of the report. It is one step in the direction of the application of risk assessment to environmental management in an orderly manner, based on common understandings and terminology.

In taking this step the report advances upon previous work in two main directions. First, it attempts to bring within the framework of risk assessment a wider range of environmental problems, especially those of accelerating the processes of deforestation, soil erosion, and desertification as well as the risks that have gained much attention in ecotoxicology.

Second, this report advances further into questions of the social evaluation of risk, and the use of risk assessment in environmental management and national policy.

This study has been made possible by the United Nations Environment Programme which has strongly supported the effort. Dr. Ashok Khosla, Director of the International Referral Service of UNEP, has been especially helpful as liaison officer, as contributor of ideas and suggestions, and as participant in meetings and informal discussions.

A major debt is also owed to the Hungarian Academy of Sciences which invited SCOPE to hold a seminar at the Research Institute for Biology, Tihany, Hungary. Their hospitality at Tihany was generous, and the facilities and support provided for the meeting in June 1977 contributed in large measure to its success. None of us who were at Tihany is likely to forget the enthusiastic support we received from our Hungarian colleagues, and our thanks are especially due to Dr. Istvan Lang, Deputy Secretary-General of the Hungarian Academy of Sciences for making the arrangements. Thirty-nine scientists from 23 countries participated in the Tihany seminar. They prepared background papers, and in a series of small working groups put together many of the ideas contained in this report. A complete list of participants and background papers is given in Appendix A. The group also worked upon an early draft of this report which had been prepared by ourselves as editors.

After Tihany the report was substantially revised and extended in the light of the discussions and reports prepared. It was not possible of course to use all the material and ideas generated, and the editorial task, of what we hope is judicious selection, has been a difficult one. The report has benefited from contributions from the following:

M. Arshad Ali Beg	Pakistan	Istvan Lang	Hungary
Guyla Bora	Hungary	T.R. Lee	UK
Rorke Bryan	Canada	Joanne Linnerooth	IIASA-IAEA
Phyllis Daly	USA	Lennart J. Lundqvist	Sweden
F.R. Farmer	UK	R.E. Munn	Canada
Richard Jelinek	Czechoslovakia	A. Oudiz	France
Y. Fukushima	Japan	Prom Panitchpakdi	Thailand
B.W. Garbrah	Ghana	E. Pattantyus	Hungary
E. Hadac	Czechoslovakia	Jan Pinowski	Poland
F.K. Hare	Canada	V. Puscariu	Roumania
G.C.N. Jayasiroya	Sri Lanka	Jorge Rabinovich	Venezuela
Tore K. Jenssen	Norway	Hussein Razavi	Iran
Robert W. Kates	USA	Ralph Richardson	USA
Homa Katourian	Iran	Hans Rieger	West Germany
P. Kazasov	Bulgaria	G. Rzhanova	USSR
Ashok Khosla	UNEP	Andrew Sors	UK
Huxley Knox-Macauley	Sierra Leone	P. Stefanovits	Hungary
		P. Stokes	Canada
V.F Krapivin	USSR	A. Szesztay	Hungary

Marta Ventilla	USA	Anne U. White	USA
G. Vida	Hungary	G.F. White	USA
M. Wasserman	Israel		

This revised report was re-examined by a small editorial group convened at the Institute for Environmental Studies, University of Toronto, in April 1978. This group consisted of Phyllis Daly (USA), Roger Kasperson (USA), Ashok Khosla (UNEP), Assefa Mehretu (Ethiopia), Timothy O'Riordan (UK), Jorge Rabinovich (Venezuela), and the editors.

The final draft of the report was circulated for comment in June 1978 to all the participants in the Tihany meeting, to the editorial group of April 1978, and to a number of other interested scientists. We wish to acknolwedge review comments and suggestions from the following: Shalom Eilati, F.R. Farmer, Frank Fenner, R.E. Munn, Edward E. Pochin, Tim O'Riordan, E. Somers and Russell Train.

Nevertheless the editors and not the reviewers or the contributors must accept responsibility for any errors of fact or judgement which remain. Nor do the views expressed necessarily reflect those of the United Nations Environment Programme or the Hungarian Academy of Sciences.

While the present report owes much to the inputs of many people, and represents a wide canvass of scientific opinion, it makes no claim to be a definitive statement. It is a step in an evolving pattern of scientific collaboration in learning to deal in a more effective manner with the kaleidoscopic problems of environmental risk.

Institute for Environmental Studies   ANNE WHYTE
University of Toronto   IAN BURTON
June 1979

Executive Summary

Environmental risk assessment involves a search for a 'best route' between social benefit and environmental risk. It is a balancing or trading-off process in which various combinations of risks are compared and evaluated against particular social or economic gains. It does not necessarily imply either a no-risk policy or even a minimum one. However, risks should be as fully understood as possible if they are to be effectively managed. Furthermore, the choice of a best route for development inevitably involves questions about the total amount of risk that is acceptable in any one area, as well as the distribution of risks among different sectors of the population.

This report presupposes that a new species of government official is emerging — the risk assessors. It sets out to introduce them to the multifaceted task of risk administration so that, in a world of increasing specialisation of knowledge, they can gain an overview of the scientific, legal, policy and management implications of risk assessment. Its aim is to help develop risk management decisions that are more systematic, more comprehensive, more accountable and more self-aware of what is involved than has often been the case so far.

The word 'risk' has been used in the literature to mean either the *probability* of danger or the hazard itself. In this report, risk means a hazard or danger with adverse, probabilistic consequences for man or for his environment. When used in 'risk assessment' the concept of risk includes not only probability and consequences but also how societies evaluate them. The risks with which this report is concerned are all in some way 'environmental'. They arise in, or are transmitted through, the air, water, soil, or biological food chains, to man. Their causes and characteristics are, however, very diverse. Some are created by man through the introduction of a new technology, product or chemical. Others, like soil erosion or natural hazards, result from natural processes which happen to interact with human activities and settlements.

There are insufficient data on the incidence and impacts of different risks to quantify their relative magnitudes and severity in the world. Even if there were such data they would not give a reliable indication of priorities on a global scale, because it is in the nature of risks and benefits that their relative values are *very differently appraised from country to country*.

The most commonly reported risks in developing countries are primarily those of resource depletion (such as loss of fauna and fish, soil erosion, overgrazing and deforestation); habitat risks (inadequate domestic water supply and sanitation); and pollution risks (air pollution and industrial waste disposal). These are all risks that can be exacerbated by the development processes of agricultural expansion, industrial development and urbanization.

The dichotomy that is sometimes implied when the population risks of high and low income societies are contrasted, can be misleading. It is increasingly becoming apparent that the assessment of all environmental risks is as urgent in developing countries as in the more heavily industrialized and higher income areas. (Chapter 1)

Modelling the problem is an important decision-making tool even where good scientific data are scarce and the model is simply a qualitative flow diagram showing causes, transmission routes and their effects. It can set the stage for asking specific questions of different agencies and enables the risk manager to keep an overall picture in view. It can also identify information needs and inconsistencies. Indeed, information breaks in the cause and effect chains of environmental models are the rule rather than the exception. Sometimes there are legal or political obstacles to filling these gaps, but even more important is the natural characteristic of the environment to vary in space and time so that it is difficult to separate 'signal' from 'noise'.

Modelling a risk system, even by a simple flow chart, immediately identifies some of the decision-points about what risks to consider in the assessment and what to exclude. For example, a decision has to be made whether risks to human health are the sole criterion or whether risks to the environment are also to be considered. A useful way to make explicit the basis on which the assessment, and eventual regulatory decisions are based, is to develop a check-list of these decisions.

Some of the decisions that need to be on such a check-list, and thereby made explicit to those managing the risks, are:

(1) Are risks from all sources included when the management of any one source is at issue? (e.g. the consideration of the amounts of lead reaching a population through the air, water, soil and biological food chains when standards for any one of them are set).

(2) What are the smallest effects to be included? (premature death, acute disease, behavioural changes, emotional effects?)

(3) What are the longest term effects to be included? (immediate damage, few weeks later, years or generations afterwards?)

(4) Is damage to the environment included and, if so, to what parts of the environment? (domestic animals, wildlife, crops, any plant, the whole ecosystem?)

The assessment process, having specified what the risks are, and which ones are to be considered in calculating the damage, usually also presents a yardstick for measuring them. Even where risks can be quantified, the figures lack meaning by themselves. This 'meaning' is usually provided by comparison of the risk under consideration with:

(1) *'Natural background levels'* of risk (e.g. flood frequency *before* massive deforestation or cosmic and background radiation before man-made sources are added).
(2) *The risk of alternatives* (e.g. different chemically based pesticides);
(3) *Other risks* prevalent in the population or region (often statistically compared in terms of probability of death or injury per exposure); or
(4) *The benefits* associated with the risks (thus higher benefits can justify higher risks). (Chapter 2)

Risk assessment includes three components. These are *risk identification* (the recognition that a hazard with definable characteristics exists); *risk estimation* (the scientific determination of the nature and level of the risks); *risk evaluation* (judgements about the acceptability, or otherwise, of risk probabilities and consequences). After the risk has been assessed, there remains the choice and implementation of intervention, or the decision *not* to intervene.

Risk identification and estimation are both concerned with collecting information on:

(1) The nature and extent of the source;
(2) The chain of events, pathways and processes that connect the cause to the effects; and
(3) The relationship between the characteristics of the impact (dose) and the types of response (effects).

In practice, risk assessment often begins by looking at one part of the problem, usually the source or the effect, rather than considering the system as a whole. This is a pragmatic response to the different ways in which risks are discovered. Once a risk is suspected, it is important to bring together as much available information as is possible before designing ways in which additional data are to be collected.

The selection of techniques requires initial decisions to be made about:

(1) The main methods to be used — monitoring, experimentation and testing, or modelling;
(2) Whether the risks arise principally out of a *technological system,* or through *environmental processes,* or through *human biology and behaviour,* because the appropriate measurement techniques differ according to where in the sequence of cause and effect you wish to measure. (Chapter 3)

In public policy, risk assessments made on the basis of scientific evidence or public alarm, have to be translated into statutes or regulations that can be enforced and, if necessary, stand up in courts of law. It is important therefore for risk managers to be sensitive to the policy and legal implications of the legislation and regulations they may propose. So far national policies have broadly taken one of two approaches:

(1) Specifying *generally* applicable codes and regulations based on what is known about the risks; and

(2) A case by case approach in which the specific circumstances of each situation allow a separate assessment to be made for each case.

The second approach is often rooted in the legal concept of *reasonableness* which has a long tradition in legal systems derived from English common law, in many traditional systems, and often underlies regulatory codes.

In environmental risk control legislation, one of the four different principles are commonly invoked. These are:

(1) *Scientific data* about cause-effect relationships which enable an 'acceptable' level of risk to be set.

(2) *No risk* acceptable at all. This is the zero exposure or zero tolerance rule exemplified by the Delaney clause about carcinogenic food additives in the USA.

(3) *The best practicable means* that can be applied to reduce risks in particular circumstances (usually with regard to cost, technical and manpower constraints and loss of benefits).

(4) *The efficacy* or effectiveness in producing benefits of the product or process that is also producing the hazard.

No matter how good the models and how reliable the estimates, there still remains the important task of deciding the *meaning* of the data collected. Decision-makers are confronted with many stories about risks, but not all of them matter. Tools have been developed to help distinguish between the more trivial risks and the serious ones. These tools include cost-benefit and risk-benefit analysis. (Chapter 4)

One outcome of a complex government machinery with different departments looking after Fisheries, Labour, Health etc. is that information becomes decentralized so that no single person or department has the necessary grasp of the whole picture. This is particularly true of information about environmental problems which fall under every department's area of interest. A way to mitigate this fragmentation is to establish a procedure for compiling a *natural risk profile,* even using, as a first step, simple actuarial data on the number and magnitude of different hazards that have occurred, together with information on their effects and when and where they took place. These data can be used to evaluate trends over time and to establish where the gaps in knowledge are that need to be urgently filled.

The organizational structures, both within and between government departments, and the nature of the links between them and the public, play important roles in risk management. Put simply, most government structures are inadequately designed to manage environmental risks. There are several ways to try to mitigate these problems; the creation of large 'super agencies'; the improvement of coordination between departments; the transformation of departments from purely functional to regional responsibilities; and the development of what are called 'matrix organizations'. (Chapter 5)

Finally, in the assessment and management of environmental risks, no nation is an island. Risk management enters into the relations between nations: some problems are transported across international boundaries by environmental processes and affect neighbouring nations, and in some cases (e.g. DDT and ozone depletion) affect the whole world. Some environmental management decisions taken in one country have repercussions in others because they are economically linked through trade or international aid programmes, or simply because of the dissemination of risk information between scientists of different countries.   (Chapter 6)

Less than a decade ago the work that is now being done by international environmental organizations like SCOPE and UNEP would have been considered utopian. Their establishment reflects a growing awareness among nations that the *common interests* of mankind in the face of growing environmental hazards outweigh national or sectoral interests. Looking to the future, we see several directions in which our collective risk management capability needs to be strengthened. These are:

(1) The development and strengthening of national risk management institutions.
(2) At the international level, further developing procedures to exchange scientific information and increase the area of agreement, and thus the prospect for international collaboration on required action.
(3) Developing mechanisms for harmonizing national decisions stemming from environmental assessment.
(4) Integrating environmental management policy more closely with international trade and development policies, particularly to reduce the danger of policies adopted in the name of environmental protection, becoming, in effect, non-tariff barriers to international trade.

From the point of view of scientific risk assessment and from the perspective offered by the present level of environmental risks, it seems only a matter of time before a truly global and urgent risk appears. The more we can get our national and international risk management houses in order, the better prepared we will be.

CHAPTER 1

Environmental Risks

1.1 WHAT IS AN ENVIRONMENTAL RISK?

The word 'risk' has two distinct meanings. It can mean in one context *a hazard* or *a danger*, that is, an exposure to mischance or peril. In the other context, risk is interpreted more narrowly to mean the *probability* or chance of suffering an adverse consequence, or of encountering some loss. Thus 'flood risk' can refer to the presence of a danger of flooding — a flood hazard, or more narrowly, a specific probability such as a 0.01 probability flood event (a 100-year flood).

Because the word 'risk' can be used in these different ways the term has led to some confusion. Three distinct views which emerged in the Tihany Workshop are recorded here because they reflect the present divergent state of informed scientific opinion.

1.1.1 Risk As Hazard

One school of thought sees risk as more or less synonymous with hazard; that is, an event or act which holds adverse *consequences*. In this view the degree of risk is related both to its probability and to the magnitude of its consequences.

1.1.2 Risk As Probability

Another school of thought would like to retain the word risk to apply solely to probabilistic statements. This school defines 'environmental risk' as the probability value of an undesirable event and its consequences that arise from a spontaneous natural origin or from a human action (physical or administrative) that is transmitted through the environment. According to this view the difference between 'impact assessment' and 'risk assessment' is that impact assessments are concerned with events that are reasonably *certain to occur,* while risk assessment is concerned with events that *may possibly occur.* Upon closer inspection the difference between 'certain' and 'probabilistic' events appears not in the *nature* of the events themselves but in the human understanding and description of the processes involved.

1

2

1.1.3 An Evolutionary View

In the third perspective the use of the term *risk assessment* is seen as an historical phenomenon. The first assessments dealing with evaluating effects or impacts did not employ probabilistic techniques. Termed *impact* assessments, they are used to describe the known impacts of various events, and employ rather straightforward quantitative techniques to estimate the magnitude of the impacts.

As the problems being assessed become more complex (which is in part due to our increased understanding of the interrelationship of events), the area of uncertainty concerning both the *nature* of the impacts and the *possibility* of occurrence became more important. To deal with these uncertainties, assessments began to make use of mathematical techniques, and in particular probabilistic theory and models of stochastic processes. Thus the assessments themselves became more sophisticated and complex. With the growing use of probability, the term risk assessment came into being to differentiate the new type of assessments from the earlier '*impact*' assessment that did not focus upon the conditional or probabilistic aspects of the event. In this view therefore risks are hazards in which the probabilistic element is important either for reasons of the state of knowledge, mode of analysis, or management, or all three.

1.1.4 Risk As Used In This Report

In this report, *risk* is taken to mean the probability times the consequence of an adverse or hazardous event. A broad meaning of risk is retained here because the report is concerned with the incorporation of risk assessment into environmental management. For the purposes of management, environmental risks have other relevant characteristics in common as well as their probabilistic nature. These characteristics justify an approach which treats environmental risks as a set of related phenomena. They include:

(1) The risks involve a complex series of cause and effect relationships. They are connected from source to impact by pathways involving environmental, technological and social variables which need to be modelled and understood in concert. There are thus common elements in the systematic approaches required for the study of risk.
(2) The risks are connected to each other. Usually several or many risks occur simultaneously within the same country, region, or city and this requires an ability to compare them and make trade-offs or balancing decisions about how much of one risk to accept in relation to another.
(3) The risks are connected to social benefits so that a reduction in one risk usually means a decline in the social benefits to be derived from accepting the risk. The social benefits of different risks are related to each other or may be very similar.

(4) The risks are widespread over the globe and concern many countries, both developed and developing. They occur in both industrial and agricultural sectors of the economy. There are advantages to nations therefore in comparing approaches to risks in the context of environmental management.

(5) The risks are not always easy to identify and sometimes identification occurs long after serious adverse consequences have been felt. There is merit in comparing the ways in which different risks arise and are recognized.

(6) The risks can never be measured precisely. Because of their probabilistic nature it is always a question of estimation. The methods for risk estimation have underlying similarities that can be described and improved.

(7) The risks are evaluated differently in social terms. Thus a risk considered serious in one place may be considered unimportant in another. It is important to understand why similar processes of risk evaluation can give rise to such dissimilar conclusions.

When used in this report 'risk' therefore means a hazard or danger with adverse, probabilistic consequences for man or his environment. When only the probability component of risk is meant, expressions such as 'risk probability' or 'probability of risk' are used. When used in 'risk assessment', the concept of risk includes not only probability and consequences but also how societies evaluate them.

1.1.5 Environmental Risks

The risks with which this report is concerned are all in some way 'environmental'. They arise in, or are transmitted through, the air, water, soil or biological food chains, to man.

Their causes and characteristics are, however, very diverse. Some are created by man through the introduction of a new technology, product or chemical, while others, such as natural hazards, result from natural processes which happen to interact with human activities and settlements. Some can be reasonably well anticipated, such as flooding in a valley or pollution from an industrial smelter. Others are wholly unsuspected effects at the time the technology or activity was developed, such as the possible effects on the earth's ozone layer of fluorocarbon sprays or nitrogen fertilizers.

While being diverse in themselves, environmental risks, as defined here, share a second common feature in addition to being transmitted through environmental media. They cause harm to people who have not voluntarily or specifically chosen to suffer their consequences, and thus they require regulation on the part of some authority above that of an individual citizen — that is, they require managing. These consequences can fall on other groups in the future as well as today, as for example in the mismanagement of natural resources. In this report, environmental risks *exclude* personal choices such as smoking, rock-climbing or tampering with electrical circuits. The immediate consequences of this latter group fall upon the individual who is voluntarily accepting such risks and the government role is usually to educate the public rather than to regulate or control the risks. Even in these cases, the transmission of risk to others through environmental media can be of concern. The risks of smoking to non-smokers present, for example, is probably small but it *is* the subject of current risk assessment and changes in public policy.

The boundaries between environmental and other risks can never be hard and fast ones and there are always marginal cases. As well as the personal risks which are excluded here, other risks are considered marginal to the central focus. These include accidents in the home, traffic accidents and food additives. While arguments can be made to include these as environmental risks they are less germane to our discussion than are risks such as soil erosion, natural hazards and water pollution.

1.1.6 Which Environmental Risks Are Important?

Many of the environmental risks that have received public attention follow on the heels of urbanization and industrialization; they are the risks of economic development. Not surprisingly these risks have been most associated with those countries, or those regions within countries, that are already highly industrialized. While it is quite possible that risks such as air pollution and toxic metals in food chains, are more severe in developed countries they are rapidly increasing in the urban-industrial regions of developing countries. Other risks are more widespread in the poorest countries — those stemming from malnutrition, inadequate housing and sanitation and the like, but they are not absent in the richer nations. Some risks — e.g. unsafe water supply — are serious in both developed and developing countries but for somewhat different reasons: contamination with small amounts of carcinogenic industrial effluents in the one case, and bacteriological contamination in the other.

There are insufficient data on the incidence and impacts of different risks to quantify their relative magnitudes and severity in the world. Even if there were such data they would not give a reliable indication of priorities on a global scale, because it is in the nature of risks and benefits that their relative values are very differently appraised from country to country. One surrogate measure for risk magnitude is expectation of life. Since expectation of life is known to be much lower in some countries than others, it may be inferred that the chief risks in those countries should be accorded some international priority.

One indication of where the important risks are thought to lie is to be gained from a list of international monitoring activities. The list in Table 1.1 therefore gives an idea of priority risk areas from the viewpoint of what it is considered

Table 1.1   International Monitoring Activities

ECOLOGICAL MONITORING

    Soil degradation — global
    Tropical forest cover
    Rangelands
    River and sediment discharge
    World Glacier Inventory
    Isotope concentration in precipitation

BIOSPHERE

    Wildlife sampling and monitoring
    Impact of pesticide residues
    Living marine resources

POLLUTANTS

    Air quality monitoring — global
    Transmission of air pollutants in Europe
    Water quality — global
    Eutrophication in inland waters
    Food and animal feed contaminants
    Pollutants in body fluids and tissues
    Human milk composition
    Pollutants in human hair
    Ionizing radiation

CLIMATE

    Climatic variability
    World Weather Watch
    Solar radiation
    Atmospheric Ozone
    Climate change
    Glacier mass balance and fluctuation
    Atmospheric pollutants

OCEANS

    Pollutants in regional seas — Mediterranean
                          — North Sea
                           — Baltic
                           — NE and NW Atlantic
    Open ocean waters
    Marine oil pollution
    River discharge to sea
    Background levels of selected pollutants

NATURAL DISASTERS

    Tropical cyclones
    Tsunami information
    Flood forecasting

Source: Martin and Sella, 1977.

6

Table 1.2   Priority Pollutants

Substances and environmental stress indicators that are potentially important with respect to their direct and indirect effects on man and the biosphere: (Munn, 1973)

1. Airborne sulphur dioxide and sulphates
2. Suspended particulate matter
3. Carbon monoxide
4. Carbon dioxide and other trace gases that affect the radiative properties of the atmosphere
5. Airborne oxides of nitrogen
6. Ozone, photochemical oxidants and reactive hydrocarbons
7. Polycyclic aromatic hydrocarbons
8. Toxic metals, especially mercury, lead and cadmium
9. Halogenated organic compounds, especially DDT and its metabolites, PCB, PCT, dieldrin and short-chain halogenated aliphatic compounds
10. Asbestos
11. Petroleum hydrocarbons
12. Toxins of biological origin (from algae, fungi, and bacteria)
13. Nitrates, nitrites and nitrosamines
14. Ammonia
15. Selected indicators of water quality: biological oxygen demand (BOD), dissolved oxygen (DO), pH, coliform bacteria
16. Selected radionuclides
17. Airborne allergens
18. Eutrophicators (e.g., nitrates and phosphates)
19. Soluble salts of the alkali metals and the alkaline earth metals
20. Other substances that have caused significant local environmental problems in the past such as arsenic, boron, elemental phosphorus, selenium, and fluoride
21. Noise
22. Waste heat

to be important to monitor on an international level. This may be compared with the list of 'priority pollutants' shown in Table 1.2. Here again a group of scientists at the international level has attempted to list substances and environmental stress indicators considered to be of priority concern.

Another approach is to ask national governments what they consider to be problems of environmental risk that affect them. A survey has been carried out by the International Union for the Conservation of Nature and Natural Resources (IUCN) in collaboration with UNEP. Sixty-three developing countries (see Table 5.1 for the complete list) were asked in which risk categories they considered problems to exist in their own country. The information produced therefore relates to the *number* of national governments

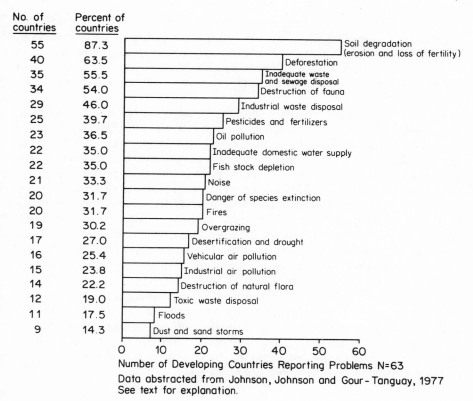

No. of countries	Percent of countries	
55	87.3	Soil degradation (erosion and loss of fertility)
40	63.5	Deforestation
35	55.5	Inadequate waste and sewage disposal
34	54.0	Destruction of fauna
29	46.0	Industrial waste disposal
25	39.7	Pesticides and fertilizers
23	36.5	Oil pollution
22	35.0	Inadequate domestic water supply
22	35.0	Fish stock depletion
21	33.3	Noise
20	31.7	Danger of species extinction
20	31.7	Fires
19	30.2	Overgrazing
17	27.0	Desertification and drought
16	25.4	Vehicular air pollution
15	23.8	Industrial air pollution
14	22.2	Destruction of natural flora
12	19.0	Toxic waste disposal
11	17.5	Floods
9	14.3	Dust and sand storms

Number of Developing Countries Reporting Problems N=63
Data abstracted from Johnson, Johnson and Gour-Tanguay, 1977
See text for explanation.

Figure 1.1 Major Environmental Risks in 63 Developing Countries
(for list of countries see Table 5.1)

recognizing these problems and not to their overall magnitude either in extent or socio-economic impact.

The twenty most frequently reported causes of environmental risk are listed in Figure 1.1. Loss of soil through erosion or depletion of fertility is reported in almost all countries, with deforestation ranking second. The most common risks therefore, in the eyes of national governments are primarily those of *resource depletion* (such as loss of fauna, fish stock depletion, soil erosion, overgrazing, deforestation and the like), habitat (inadequate waste and sewage disposal, domestic water supply), and *pollution* risks (air pollution, water pollution by oil and industrial and toxic waste disposal).

These are all risks that can be exacerbated by the development processes of agricultural expansion, industrial development, and expanding populations in cities and on the land.

Water pollution is by implication a frequent subject of concern in developing countries where untreated industrial waste and urban sewage are dumped into rivers that form the water supply for people living downstream. In addition, in urban areas, major risks are seen to be air pollution from traffic and from industry, together with noise. In rural areas, habitat risks of inadequate and polluted water supply together with poor sanitation are frequently reported.

The low frequency with which natural hazards appear merits comment. Only floods, drought and sand storms occur in the first twenty causes of environmental risk. Tropical cyclones and earthquakes are conspicuously absent. This is possibly because national governments see such risks as less amenable to management or national policy decisions. It probably does not mean that natural hazards are *not* considered to be major environmental risks.

A major caveat must be entered against the lists of international monitoring (Table 1.1), priority pollutants (Table 1.2) and major risks (Figure 1.1) What emerges on these lists depends on the specific purposes for which they were designed, the people asked, and the way the questions were put. The lists are therefore *indicative* of a range of problems, but are not exhaustive, nor 'objective' in rank-ordering the risks.

These data do suggest that environmental risk management in developing countries is now, and will increasingly be concerned with the chemical pollution of land, air and water through industry, urbanisation and the agricultural use of chemicals. This concern does not exclude the problems of basic needs — adequate provision of food, water and shelter — but is additional to them and occurring simultaneously.

1.2 TYPES OF ENVIRONMENTAL RISKS

In the context of governmental management the many different environmental risks can be considered as falling within one of five principal domains. These are public health, natural resource management, economic development, man-induced and natural disasters, and the introduction of processed food, drugs and consumer products. The risk assessment process has different characteristics in each of these domains.

1.2.1 Public Health

In many countries, chronically poor health caused by malnutrition, parasitic and infectious diseases, food and water contamination, inadequate sanitation and health services and poor housing is a main area of concern for national governments. Where public health risks are high, other environmental risks such as ozone-layer depletion or DDT induced loss of birds of prey, are usually regarded as less urgent. In any country, the importance attached to other environmental risks is influenced by the need to improve human health, so that the domain of public health risks can be considered as the baseline against which the costs and benefits of other risk reducing programmes are inevitably evaluated.

1.2.2 Natural Resources

Resource management is often concerned with controlling environmental risks such as soil depletion and deforestation which are caused by existing practices (or mismanagement), particularly in the primary economic activity sector. Subsistence agriculture (including migratory livestock herding) can create environmental stresses and resource depletion. A risk management policy may have to deal with a large number of agriculturalists in whose minds the question of risk to nature or to others is not associated with their traditional practices and who themselves have little resources to undertake mitigating measures. The problem is often widespread, particularly in the less accessible rural areas, and there is usually little centralized administrative machinery to impose regulatory measures.

By virtue of the magnitude of the problem of environmental degradation caused by mismanagement of natural resources (especially of soil and forest caused by subsistence (traditional) farming) very few governments can develop a capacity in the near future to cope with that problem to a satisfactory degree. Nor does the population affected by this risk, which is a large part of the population in virtually all countries, have the perception to demand appropriate government response for soil erosion or deforestation.

1.2.3 Economic Development

A different risk management problem is associated with environmental stress brought about by planned and deliberate new forms of economic activities which, from the outset, are often known to have inherent characteristics in producing environmental risks. In the context of development processes that take place within the framework of national or regional planning, the government bears virtually the entire responsibility for making sure that the risks from induced development processes do not exceed acceptable levels. Moreover where the government deliberately introduces a series of development projects such as mining, irrigation schemes, manufacturing industries and the like, then the problem of environmental risk becomes more focused and defined for the decision makers to institute appropriate safeguards and regulatory mechanisms to have some level of control on the environmental risk that might ensue. These safeguards can be built into the development process.

1.2.4 Man-induced And Natural Disasters

Where the environment is subject to natural catastrophies, the decision maker's role has tended to be restricted by and large to *post facto* corrective measures and rehabilitation. More recently, especially as a result of lessons from the Sahelian experience, for example, there is a pressure on decision makers to look into early warning systems which may be used to reduce the effects of environmental hazards and to institute damage reducing measures such as land use zoning and appropriate building regulations. In many areas where disasters strike, the local resources are too poor to pay for these improvements and the task falls to the national government, either alone or with the help of bilateral and international aid and disaster relief programmes.

1.2.5 Introduction of New Products

Perhaps one of the more well known risk management systems in any country is that designed to regulate the introduction into the market of processed food and drugs, chemicals such as pesticides, and consumer products, of both local and imported origin. Most countries have a regulatory system which requires sampling for quality control and toxicity, and has provisions for initiating legal proceedings against offenders. The risk assessment problem is usually one of inadequate facilities for the size of the task, whether in an industrial country or in one of the least developed nations. More products, drugs and chemicals are introduced into countries than they can adequately test, monitor and regulate and significant time lags develop between the introduction of new products and the assessment of their risks. The management task is generally one of trying to increase the scientific and administrative resources to keep pace with the need for regulation.

The problem is exacerbated in some fields by the speed with which products are dropped and replaced by new ones which are only slightly different. This is notoriously the case with drugs, which are frequently replaced by new products by the time regulatory tests have 'caught up'.

1.3 THE MANAGEMENT OF RISKS

Environmental risk management involves the search for a 'best route' between social benefit and environmental risk. It is a balancing or trading-off process in which various combinations of risks are compared and evaluated against particular social or economic gains.

In a previous volume in this series by Kates *(Risk Assessment of Environmental Hazard: SCOPE 8)* risk assessment was described as having three interrelated components: *risk identification, risk estimation* and *risk evaluation*. This volume follows Kates' nomenclature and uses his work as a starting point for a discussion of questions surrounding the implementation of a risk assessment approach.

1.3.1 Risk Identification

Risk identification simply means recognizing that a hazard exists and trying to define its characteristics. Often risks exist and are even measured for some time before their adverse consequences are recognized. In other cases, risk identification is a deliberate procedure to review, and it is hoped anticipate, possible hazards.

1.3.2 Risk Estimation

This is the scientific determination of the characteristics of risks, usually in as quantitative a way as possible. These include the magnitude, spatial scale, duration and intensity of adverse consequences and their associated probabilities as well as a description of the cause and effect links. Both risk estimation and identification can involve modelling, monitoring, screening and diagnosis (Kates, 1978, pp. 14-19) which are discussed in Chapters 2 and 3 of this report. The main purpose of these two management functions is to understand the environmental system and its complex pathways and processes through which risks occur.

1.3.3 Risk Evaluation

The third component of risk assessment is *risk evaluation* in which judgements are made about the significance and acceptability of risk probabilities and consequences. This stage is central to policy determination. Evaluation techniques seek to compare risks against one another, and against benefits, as well as providing ways in which the social acceptability of risks can be judged. Indeed, any judgement about social acceptability combines both political and managerial decisions since it inevitably involves a calculation of who is likely to benefit and who to suffer, and what compensation, if any, should be paid.

After a risk has been identified, estimated or evaluated (or any combination of the three) there comes a point where some kind of intervention (or deliberate decision *not* to intervene or to delay action) takes place. The nature of the intervention varies greatly depending not only on what the risks are (and are perceived to be) but upon the particular policymaking 'style' and the constitutional and administrative framework. But before that point of implementation has been reached, a great deal of risk assessment has already taken place, and has profoundly affected the course of events that will follow.

1.4 WHY WE NEED ENVIRONMENTAL RISK MANAGEMENT

The focus of scientific research on problems of the environment has highlighted many gaps and inadequacies in present knowledge. The pressure of events requires, however, that important decisions about environment and development be made now rather than at some indefinite time in the future. To do so involves making decisions under conditions of risk and uncertainty. The concept of risk has therefore become central to the environmental management process. How can a course of development be chosen which is 'safe enough'? A safe enough, or less risky, course of development would be one which would avoid the dangers of collapse through unsupportable or unsustainable development. In other words, it would be development compatible with the environment — or ecodevelopment. It would also minimize or reduce to acceptable levels undesirable side effects, for those subject to risk, but also for those who create risks and those responsible for managing them.

The choice of a 'best path' for development involves not only questions about the total amount of risk that is acceptable in any one area, but also the distribution of risks among the population. Thus risk management is central to the ecodevelopment process in two ways. First, it is necessary to ensure that the risks taken will not undermine or negate the aims of development. Second, both the benefits and the risks should be distributed in a socially acceptable way.

Societies differ widely in the spectrum of risks that they encounter and in their view of the priorities to be favoured in dealing with them. In some countries there is major concern over cancer, birth defects and mutations and their possible causes in man-made and man-modified environments. Elsewhere the societal priorities are more centered on those risks associated with the lack of basic needs — water that is safe to drink, housing and nutrition that permit the healthy growth of individuals, families, and the community, and the development of natural resources that does not result in the irreversible destruction of soil, forests and wildlife.

The dichotomy that is sometimes implied when the population risks of high and low income societies are set in contrast, can often be misleading. It is increasingly becoming apparent that the assessment of all environmental risks is as serious and important a responsibility in developing countries as in the more heavily industrialized and higher income nations.

Indeed, countries now undergoing rapid industrial development or large scale expansion of commercial agriculture, are confronted by an especially difficult situation. They combine in a demanding fashion some of the traditional risks of natural hazards and resource depletion with the new pollution and technological hazards associated with industrial development and modern agriculture. Paradoxically, the more successful the economic development process, the more likely there is to be generation of new risks at the same time that unprecedented pressures are arising in the more 'traditional' risk areas of soil erosion, deforestation, desertification and natural hazards.

Furthermore, as rapidly developing nations are drawn more strongly into the pattern of international trade and commodity flow they find that standards and regulations established elsewhere for the protection of the environment and human health can have a deep and lasting effect on their development. Sometimes these regulations are appropriate to their needs, but often they are not.

Environmental risk management therefore raises questions for all nations, both in their own internal or domestic affairs and in relation to others in the family of nations. This report is not therefore addressed exclusively to one group of nations or another. It attempts to elucidate the problems of environmental risk assessment especially in its international dimensions, and to show how it relates and fits into decision-making in economic development.

There is a great deal of scientific information about some environmental risks. This originates largely from countries rich in scientific and technical manpower and from research institutes established to look closely into environmental risks. In addition, international organizations draw upon this wealth of scientific information to set or suggest international standards and guidelines. It is not wise, however, for a national government to assume that because a risk has been identified, assessed, and a standard established in one or more countries, that this evaluation will automatically apply to their own country. The *consequences* of risk vary from place to place, both as measured in scientific ways and as perceived by local populations. However, when scientific manpower, management skills, and institutional capability are in short supply, it may be a misallocation of resources to invest a large effort into research on the toxicity of industrial effluents or the ecological affects of pesticides, if this has been done elsewhere. Independent risk evaluation does not necessarily require replication of all the scientific work required for risk estimation.

The management of risks does require resources — money, skilled manpower, and time — and is itself associated with the risks of cost, delay and inaction. Risk management is not, however, an entirely new or unfamiliar exercise. Governments already weigh the risks of the exhaustion or depletion of a fishery while building new fishing boats, and farmers have long appraised the risk of a drought or a pest infestation while planting a crop.

Environmental risk management is only part of a much larger governmental set of national needs and priorities. Social and economic development often lead to the introduction of new processes and products, and to the development of hazard prone areas without any consultation with risk assessors or environmental scientists. The environmental risk manager, whether at a high Ministerial level or as an individual technician, has to compete with other demands in a nation's resources and attention. Often he will be faced with risks whose origins lies deep in social customs and history, and thus cannot be improved without more far-reaching changes than can be encompassed by environmental management alone.

It is not the purpose of this report to suggest that no risks be taken. However, they should be understood as fully as possible. This means that the factors that are taken into account in any decision need to be expressed clearly and where appropriate in terms of the risks involved. Consequences need to be explained and understood both by the authorities and by those at risk, before they are (knowingly) accepted. Only in this direction lies the way to more effective risk management and to a safer and more prosperous future.

CHAPTER 2

Establishing an Overview of the Problem

2.1 MODELS, BOUNDARIES AND CONTEXTS

The soil from one field is washed down a gully and deposited as silt in a river bed downstream. In the process it has contributed to a loss in food production and an increase in flood hazards. Elsewhere, a toxic substance is emitted from a factory waste pipe into a river. It leads to contamination of local fish and a minute amount of the toxic waste finds its way into the water supply of a city downstream. To the benefits of eating fish or drinking water must now be added the risk from consuming the toxic substance.

We may aggregate many such small and localized examples, and so build up a model of the ramifications of these effects for a river basin, an ecosystem or an economic system. At the low end the impacts may be so small or so uncertain that most people would agree that they should not be a cause for concern. At the other end of the scale, where enough small and marginal changes are added together, an obvious environmental risk may be created. Between these two points there exists a large grey area where different people, organizations and nations arrive at quite different views of the nature of the risk assessment problem, and therefore of what a risk manager needs to know.

Commonly these different perceptions arise from the constraints of incomplete and often inadequate scientific knowledge of the impacts, or from the organizational limitations of their scientific and administrative departments. Funds to collect data, to make observations or tests, may be too limited, or the area of their responsibility may be confined to individual components of the problem such as soil erosion or drinking water quality, so that they cannot easily enquire into the links and interactions in a chain of events.

The approach taken here is to discuss the development of the risk assessment formulation in terms of three decision-making areas:

(1) SYSTEMS MODELLING — providing as comprehensive and detailed a picture of the problem area as possible, quantified as far as possible and with probability estimates where needed.
(2) SETTING BOUNDARIES — defining the limits of the problem beyond which other data can be ignored or neglected for decision-making purposes.

15

(3) PROVIDING CONTEXT — putting the problem assessed into a larger context by considering other risks and/or benefits.

Approached in this way, the translation of scientific data about potential effects into regulatory action to control effects, can be a more systematic and accountable procedure than is the case in the *ad hoc* and irrational process that it sometimes appears to be. To achieve rationality and accountability it is important to identify clearly and make explicit both the points of scientific uncertainties and the points of judgement that go beyond available scientific evidence.

2.2 MODELLING THE PROBLEM

Some examples of environmental risks that have been partially modelled either qualitatively or quantitively are given in Chapter 3. Relatively few comprehensive and quantitative models of environmental risks exist at present, but it is an area of much current scientific research. Even where good scientific data are scarce, a qualitative flow diagram of possible sources and pathways for a particular effect can be a valuable decision-making tool. It can set the stage for asking specific questions of different agencies and enables the decision-maker to keep an overall picture in view. It can also identify information needs and inconsistencies.

Take, for example, a case of a chemical pollutant such as a heavy metal which is discharged into the air and water by two neighbouring factories located on the same river. Figure 2.1 illustrates a generalized model of the pathways through which the pollutant is transmitted into the environment and up the food chains to man.

In an ideal world, scientific data would be available to the decision-maker on the amounts of contamination reaching each component (fish, grazing animal, milk etc.); the amounts stored or transmitted up the food chain over time; and the significance of these amounts in terms of short term and long term harmful effects. The risk assessment could follow a series of measurements and calculations like those shown in Figure 2.1 and arrive at regulations and standards aimed at controlling the pollutant at various points in the risk system. Furthermore, each standard would ideally be arrived at in the context of the total risk model — that is, all the sources and pathways combined. Advanced technological risk assessments such as for nuclear reactors or aircraft go through similar procedures.

The decision-maker is rarely, if ever, faced with such a comprehensive or detailed picture of what is going on, even for pollutants such as lead, which have been with us for centuries, or for technological risks subjected to much analysis. The task is usually one of piecing together isolated sets of data which refer to different components of the system.

Information breaks in the cause and effect chains of environmental models are the rule rather than the exception. Often there are specific barriers to obtaining information to fill these gaps. For example, in Figure 2.1, government agencies may be able to monitor the river water quality below both factories but be prevented by legislation protecting the confidentiality of industrial processes and formulae from actually measuring the concentration of pollutants in the outflow pipes of each factory. Similarly, public agencies can often not monitor air emissions within industrial premises. The agencies may also have divided responsibility for monitoring industry, as often happens between Health and Labour agencies. Technically, the levels of contaminants in the air or water may be too low to measure even though they may pose a risk to human health through biomagnification up food chains.

Sometimes concentrations of the pollutant in a food source may be known but not how often and how much people eat that food. The deposition of an airborne pollutant can be measured directly using filter instruments, but monitoring of the transmission of the deposited pollutant through the soil and ground water systems is more difficult. Yet without an understanding of this transport and storage, long term contamination of fields and ground water is hard to evaluate. These examples can be multiplied for every environmental system.

Figure 2.1 is a generalized model and does not show the uncertainties and decisions that take place within each 'box' and which compound the managerial problem. The problems associated with environmental monitoring systems for measuring air and water quality illustrate the complexity of each component within the system.

2.2.1 Environmental Variability

A characteristic of the environment is its natural variability in space and time, making it difficult to separate 'signal' from 'noise'. Even under the best of conditions, there are problems in detecting the early stages of an impending environmental hazard. For example, a depletion of stratospheric ozone would cause an increase in the ultraviolet radiation reaching the earth's surface and in the incidence of skin cancer. However, because the day-to-day, month-to-month and year-to-year variabilities in stratospheric ozone are very large, there would be considerable difficulty in detecting long-term trends. In fact, Pittock (1974) has suggested that if a 2% depletion of stratospheric ozone were suddenly to occur, an additional 10 years of observations would be required before the event could be confirmed with 95% confidence.

2.2.2 Uncertainty in Knowing What to Monitor

There is frequently difficulty in deciding what to monitor, for one or more of the following reasons:

(1) The environmental effects of a pollutant often depend not only on the toxicity and concentration of the substance but also on a multitude of associated factors, e.g., temperature, humidity, the coexistence of

18

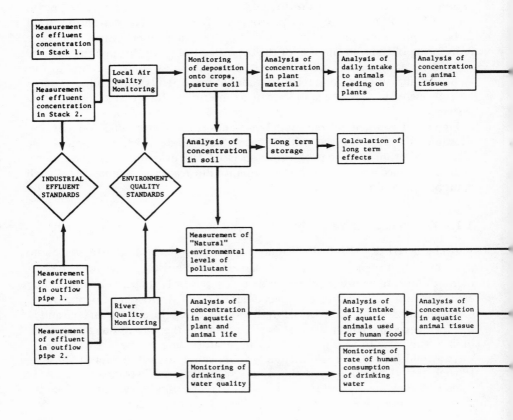

Figure 2.1 Generalized model of envirnomental pathways for an industrial pollutant, and points of entry for regulation

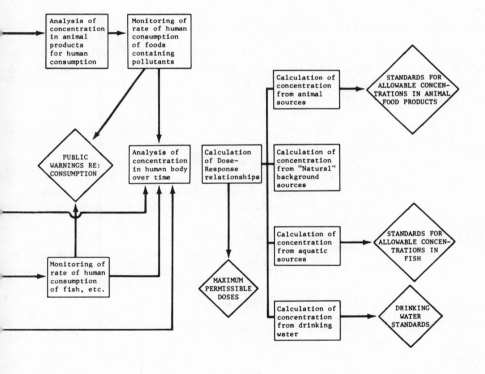

other substances, etc. Particularly in the case of suspected synergisms, existing information on dose-response relations may be insufficient to select the variables to be monitored.

(2) Sometimes there is uncertainty concerning whether to monitor the environmental doses, the biological responses, or both.

(3) In order to devise management strategies, information on the pathways of a pollutant from source to receptor is often required. This implies that a large number of associated variables should be monitored.

(4) A management strategy may be suitable for reducing a particular hazard, but at the risk of creating another hazard, e.g., replacing an air pollution problem by a water pollution one. It is sometimes difficult in such cases to decide on an appropriate monitoring design.

2.2.3 Uncertainty in Knowing Where to Monitor

There are spatial gradients (rates of change) in the concentration of pollutants, in the populations of living organisms, in the values of climatic elements, and so forth. These gradients lead to the following problems in the design of monitoring systems:

(1) Is it better to monitor in regions of weak or of sharp gradients? It is usually preferable to monitor at locations where environmental stresses (pollution concentrations, temperatures, etc.) are highest and where effects are most likely to occur. However, the spatial gradients are usually largest at such locations, e.g., close to a chimney, at the edge of a forest stand, etc., making it difficult to obtain a representative reproducible sample.

(2) Even in regions with weak spatial gradients, there are many random local variations. Should a 'typical' site be selected, or one in which these local irregularities are minimized?

2.2.4 Uncertainty in Knowing When to Monitor

The records of environmental elements and indicators are time series in which each observation is correlated with its predecessor. Wherever possible, therefore, continuous monitoring should be undertaken to avoid the biases introduced by irregular sampling, although the higher costs of continuous monitoring may not make it possible.

2.2.5 Uncertainty in Knowing How to Monitor

The perception of an environmental hazard is sometimes a function of the lower limits of detection of a pollutant or of a resulting effect. As analytical methods and sensing devices become more sensitive, the range of problems for study may, therefore, widen.

In many cases, the environmental concentrations of pollutants believed to be associated with significant effects are measured in parts per million or parts per billion. Considerable care is then required to reduce experimental errors to such a level that meaningful environmental monitoring data can be obtained.

Each of these decisions can affect the data or their evaluation and can therefore be matters of scientific or political dispute. Frequently difficulties arise in calibrating the amount or significance of one measured concentration against another, when either different methods or different parts of the system are involved. Experimental and epidemiological evidence can appear to be contradictory (such as is often the case with potential carcinogens). When these sources of uncertainty are added up across a sequence of events in a system, the task of the decision-maker in assessing the risks is clearly a judgemental one.

2.3 SETTING BOUNDARIES TO THE RISK SYSTEM

The drawing of a flow chart of the risk system immediately identifies some of the decision-points about what risks to consider in the assessment and what to exclude. For example, a decision has to be made whether risks to human health are the sole or major criteria for assessment, or whether risks to domestic animals, wildlife, crops or any part of the environment are also to be considered. Some questions about boundaries which commonly arise in risk assessments are discussed below. A useful decision-making tool to help make explicit the basis on which the assessment and eventual regulatory decisions are based is to develop a check-list of these decisions.

2.3.1 Are Risks From All Sources Included?

A farmer may lose crops and income through a variety of natural and man-induced hazards — drought, floods, landslides, soil erosion, pests, soil salinization and gullying. Equally important may be economic factors beyond his control — a worldwide glut of one commodity and a fall in prices, a ban on his product through its contamination by high residues of pesticides or toxic heavy metals. From which of these risks will the farmer be protected or compensated by government sponsored subsidies, insurance or disaster relief? And when assessing the riskiness of any economic activity in an area, either with a view to encouraging regional development or to controlling it (e.g. on a flood plain) through land use regulations, exactly what natural and socio-economic risks are to play a part in the calculations?

A child may suffer lead poisoning from ingesting lead in drinking water from lead pipes, from paint or soil containing lead, from food contaminated from lead in the air, water, or soil, or from biological food chain linkages. When setting ambient air standards or permissible industrial effluent standards for lead, are the risks of lead reaching the child from *all* other sources to be included?

2.3.2 What Are The Smallest Effects to be Included?

The most unequivocal effect of a risk on human well-being is death of an individual. For other animals and for plants, it can be the death not of any single individual, but of sufficient numbers to threaten a resource base, an ecological community or a species that is the threshold for concern. Thus the death of countless fish becomes a starting point for risk assessment and regulatory measures when fish stocks are severely depleted. Similarly hunting of whales in the Pacific or wild game in East Africa lead to governmental risk assessments when the continuance of the species in the area is threatened.

Within human health and well-being a hierarchy of effects can be identified from acute illness and death through chronic and long continued diseases which impair life, to minor and temporary ailments. Even lower down this scale would be temporary emotional effects or behavioural changes (Table 2.1).

Table 2.1   Range of Direct Risk
Consequences on Human Health

Premature Death of Many Individuals
Premature Death of an Individual
Severe Acute Illness or Major Disability
Chronic Debilitating Disease
Minor Disability
Temporary Minor Illness
Discomfort
Behavioural Changes
Temporary Emotional Effects
Minor Physiological Change

Particular problems of assessment arise with chronic diseases which are often difficult to relate to specific risks or sources of risks. Respiratory diseases for example may be caused by any or all of: occupational exposure to dust at work; air pollution from industrial and urban sources; or smoke pollution within the home; or by increased susceptibility of individuals. Chronic diseases, caused by contaminants such as asbestos, also have very long latency periods of 30-40 years so that the origin of the cause and its effect may be widely separated in time, and the individual concerned may have moved to another place and be in another occupation.

Other problems arise with effects that are considered 'sub-clinical' or not detectable upon physiological examination. The significance of other physiological effects is uncertain. For example, it is known that at blood lead levels *below* 100 micrograms of lead per 100 ml. of whole blood, clinical anaemia does not usually occur. However lead at these low levels in the blood affects the activity of one of the enzymes (ALA dehydrase) in the haemoglobin chain (Hernberg, 1972). Is this a significant effect on human health? Scientists and national governments differ in their assessments.

2.3.3 What Are The Longest Term Effects to be Included?

The effects of environmental risks may be seen within seconds, or may take years, or even generations to be revealed. The management problem lies not at the short term end of the scale, where death from drowning in a flood, or injury through a traffic accident are immediate, direct and unequivocal effects of visible causes. Rather the problem lies in deciding which future long term effects are reasonable ones to include in an assessment.

For example, overgrazing or overcutting in an area may lead to deforestation and soil erosion some decades later. Yet these effects have often been knowingly accepted for shorter term gains of economic productivity, or by necessity. To take another example, the practice of keeping children indoors in smoke-filled homes in some cultures is thought to lead to a high incidence of bronchitis and other respiratory diseases later in life. And as has been discussed before, many carcinogens seem to take as long as 30-40 years before any ill effects are seen. The question is, how far can these future effects be incorporated in present decisions and practices?

These long latency periods for carcinogens and other diseases make it very difficult to determine the cut-off point for the assessment of future risks. Even more so do the teratogenic (inherited birth defects) and mutagenic (adverse mutations occurring in germinal cells) risks of synthetic chemicals. These risks involve several generations and thus often lie in the realm of the hypothetical rather than the demonstrable. This futurist element is compounded by the low probability and scientific uncertainty that is associated with many long term risks.

The biological effects of radiation has been one of the most exhaustively assessed risks of modern technology. Even so, the assessment of genetic effects of low-level radiation on experimental mice is far from definitive:

> Experiments performed at high radiation levels show that the dose required to double the spontaneous mutation rate in mice is 30 roentgens of x-rays. Thus, if the genetic response to x-radiation is linear, then a dose of 150 millirads would increase the spontaneous mutation rate in mice by ½ per cent. *To determine by experiment whether 150 millirads will increase the mutation rate by ½ per cent at a 95 per cent level of confidence requires a test population of 8,000,000,000 (8 billion) mice.* A reduction in confidence level to 60 per cent would still require 195 million mice. (Weinberg, 1972)

Statistical low probability models, laboratory testing over several generations of animals, and the collection of detailed epidemiological and health data for occupational and general human populations are some of the tools required for assessing long term effects.

Where people are under the threat of immediate and direct risks such as inadequate water supply or starvation, the longer term risks of carcinogenesis from chemicals in the water or food are less likely to be included in the risk assessment because they are further in the future and more uncertain. As the direct health risks decline in importance within a population, concern with longer term and more indirect risks increases and at the same time, data about these long term risks become available with improved monitoring and testing procedures. The evolution of risk assessment has therefore been towards including longer and longer term effects.

2.3.4 Are Risks to the Environment, other than to Man, Included?

In the past risk assessments have tended to focus on human health but increasingly, hazards to animals, plants and natural areas have become a cause for governmental action in their own right. Many countries, both developing and developed have enacted legislation to protect natural areas from further encroachment of man's activities (Table 5.4). Similarly, many individual species of animals and plants are protected in different countries. In the USA, it has been suggested that elements of the environment such as trees should have similar legal rights to man, and be able to be represented in court (Stone, 1974).

The inclusion explicitly in the risk assessment of harmful effects other than on human health allows much more evidence to be introduced on the 'risk' side of any equation. Data on genetic effects are more readily obtainable for animal and plant populations with shorter life spans. The inability of plants and animals to sue man for damages under present legal systems has allowed both more damage to be perpetrated and more information about the effects to be gathered. Ethical and legal considerations have been more often

Table 2.2   Harmful Effects Specifically Considered in National Pesticide Legislation

	HARMFUL EFFECTS ON:	COUNTRY
MAN	— human health	Finland
MAN, ANIMALS	— human health, warm blooded animals	Turkey
	— human health, bees	Austria
	— human health, domestic animals, bees	Denmark
	— human health, mammals, aquatic animals	Japan
	— human health, cattle, wildlife	Portugal
	— human health, wildlife, aquatic animals	India
	— human health, domestic animals, wildlife	UK
MAN, ANIMALS, PLANTS	— human health, domestic animals, beneficial insects, wildlife and domestic plants	Sweden
	— human health, bees, animals, plants	Switzerland
	— human health, animals, crops	Korea
MAN, ECOSYSTEM	— human health and the environment	Spain
	— human health and the environment	Netherlands
	— human health and the environment	France
	— human health and the environment	Ireland
	— human health and the ecosystem	New Zealand

Sources: OECD, 1971; Mootooka, 1977.

implicated in the gathering and publication of human health data, particularly where man-made risks are involved. Thus the extension of the risk assessment to consideration of environmental impacts is likely to produce more data, much of them conflicting, and involve more scientists and specialized agencies in the assessment process. It is also more likely to arrive at assessments of unacceptable risks.

Table 2.2 shows the range of environmental risks considered within pesticide legislation and gives examples of countries using these terms of reference. The range varies from a narrow focus on impacts on human health only, through various inclusions of domestic animals, wildlife and beneficial insects, to crops and plants, and finally to consideration of any harmful effects in the environment. This range can also be regarded as a historical trend in the evolution of environmental risk assessment in national policies. It is also witness to the increasing complexity of the assessment task.

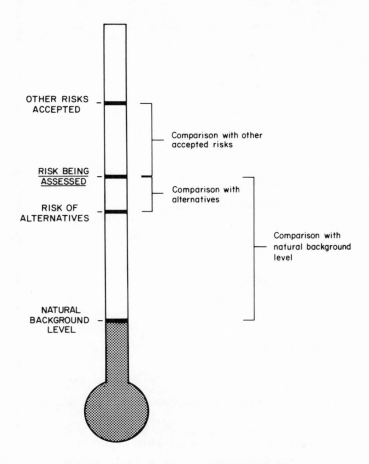

Figure 2.2 Alternative yardsticks for measuring risk
(these are not necessarily always in this order)

2.4 PUTTING THE RISK INTO COMPARATIVE CONTEXT

The assessment process, having specified what the risks are, and which ones are to be considered in calculating the total risk, usually moves to a third stage — that of providing a yardstick for measuring the risks. Even where the risks can be quantified in some way; for example 1.5 mg fluoride per litre in drinking water or 10^{-6} probability of an earthquake of a certain magnitude; these figures lack meaning when taken out of context. Risks need to be compared or measured against some yardstick to become relevant to decision-making. Three commonly used yardsticks are 'natural back-ground' levels; the risks of

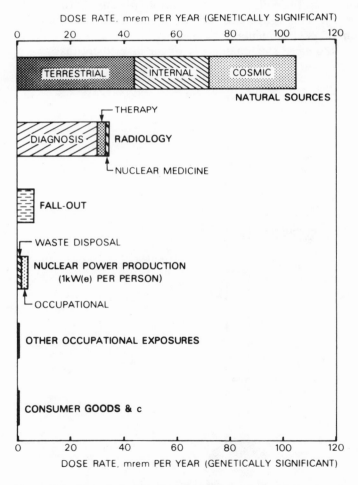

Figure 2.3 Annual genetically significant dose rate for low level radiation, averaged through whole population (Canada)

Source: Aiken, Harrison, and Hare, 1977. (Reproduced by permission of the Ministry of Energy, Mines and Resources, Canada)

alternatives; and the dangers of quite other hazards that are not strictly alternatives.

Figure 2.2 shows how these yardsticks can differ in the answers they give, for a given level of a particular risk. In the illustration, the risk being assessed may be elevated above natural levels, and be more noxious than its alternatives, but be acceptable because it is less dangerous than other risky activities or products people accept. The levels are not always in the same relative positions. In the case of radiation, for example, natural background levels are higher than either radiation leaks from nuclear reactors, or than the total radiation burden of other risks such as medical x-rays (see Figure 2.3).

These three ways of comparing the risk being assessed to other risks tend either to ignore the different benefits of alternatives, or to make them equivalent for the purposes of calculation. A fourth, and most commonly used approach is to compare the risks with the benefits, focussing either on the issue at hand, or enlarging the scope to consider the risks and benefits of alternatives. In this way, different bases for comparing risks lead to quite different 'risk equations'. These are illustrated in Figure 2.4 where the risk being assessed is shown in the centre and the alternative ways of comparing it to something else outside the risk system itself are shown as different directions. The resulting risk equations can be termed:

Elevated Risk
Balanced Risk
Comparative Risk
Risk-Benefit

In turn these different approaches involve different analytical methods.

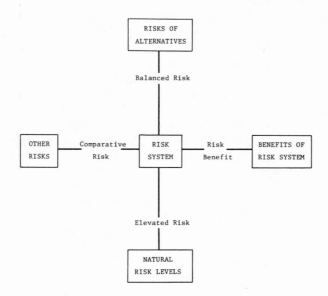

Figure 2.4 Different ways of comparing a risk

2.4.1 Elevated Risk: Comparison with Background Levels

Risks can be measured by asking what is the *additional* hazard they present over

(1) What occurs in the natural environment
(2) What has been tolerated for long periods of time without apparent ill-effects
(3) What is believed to be a beneficial amount

The definition of 'natural' environmental levels or conditions is a difficult one to determine. It depends on baseline studies prior to the source of risk being assessed, or a large number of measurements in affected and non-affected ('natural') areas. Determination of the additional flooding risk on the Indus Plain of Pakistan caused by deforestation in the Himalayas relied on a comparison of floods in the last twenty-five years (since deforestation has escalated) with records for the previous sixty-five years (Leiftinck, Sadove and Creyke, 1969). Baseline data on the amount of cosmic radiation received and radiation levels in rocks and soil have been obtained from many sites around the world so that the 'natural' radiation levels and their variability are well documented.

For some pollutants, such as lead, baseline data on natural levels in man prior to industrial or traffic sources are difficult to obtain because they have been present for a long time. Comparison of levels found in urban (affected) and rural (relatively unaffected) areas have been used to determine the lowest (and hence 'normal') lead levels in human blood. This is less than 1% or 10-30 micrograms of lead per 100 grams of blood for adults. Thus anything above this may be considered an elevated level of risk although it may not produce significant effects on human health until much higher levels are produced (symptoms of clinical lead poisoning in adults appear at about 80 micrograms per 100 grams of blood). There exists a considerable range between the lowest, 'natural' levels and levels so elevated that obvious ill-effects are seen. It is on the significance of elevated, but not excessively high levels, that much scientific and thus, regulatory uncertainty lies. As concentrations get higher, not only the percentage of the population affected increases, but also the severity of the effects.

Natural background levels have been used as a yardstick to measure the additional risk of radiation from nuclear power production (Figure 2.3) the potential risk of adding fluoride to domestic water supply as a public health measure, and the assessment of elevated noise levels near airports and traffic routes. One caution is appropriate here. There may be undetected ill-effects from background levels or long-tolerated levels. For example, background 'natural' radiation in Kerala, India, is up to ten times what is considered normal elsewhere, but medical statistics are too poor to detect a substantial harm rate.

2.4.2 Balanced Risk: Comparison of Alternatives

Another way to make a risk level meaningful is to ask what the alternatives are. These may be alternative products, processes or courses of action. This approach is best suited to situations where the alternatives being weighed against one another are indeed alternatives — they can substitute for one another to provide the same goods or benefits. Take, for example, the case of pesticides, where several alternatives may be available to protect crops against pests. When the risks of a particular pesticide, such as dieldrin, were assessed by the EPA in the United States in 1970 as too high for continued use in the USA, one of the issues in the five years of argument and litigation that followed was the risks of alternatives, especially heptachlor and chlordane. In fact, these risks were uncertain as the alternative pesticides had not yet been assessed by EPA, and eventually the EPA ban on aldrin and dieldrin was upheld.

The risks of asbestos as an insulating material has also been compared to the risks of other forms of insulation, such as fibreglass products which have similar physical properties of minute fibres, and may therefore have comparable effects on human health. The comparison of risks of alternative water supplies in a rural area where no supply is 'safe' can similarly make meaningful in a larger, more practical context, the level of contamination of a particular stream.

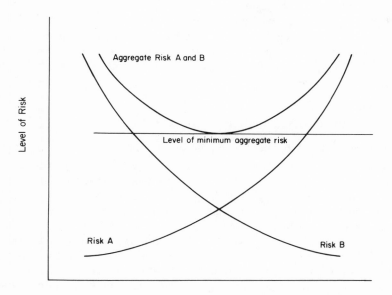

Figure 2.5 Balanced risk model

The concept of balanced risk is founded on the principle that it does not make sense to reduce a risk beyond the point at which an optimal balance of risks is attained. Consider the two risks A and B in Figure 2.5. As Risk B declines (e.g. risk of soil erosion from deforestation), Risk A increases (e.g. loss of subsistence farming production). If these risks are connected, as is commonly the case, then a balancing point (optimal level) exists in theory, beyond which an aggregate of the two risks increases in either direction. The technique of balanced risk is therefore to find that optimal level for each risk which results in the *lowest aggregate risk*. This concept need not simply be applied to risks that are causally connected. Since all risk reduction involves some loss (resources must be allocated to reduce the risk) it can be argued that resources should be allocated on a priority basis to reduce the largest marginal risks, e.g. to buy the maximum increment in safety for a given investment.

Such calculations are, of course, impracticable in quantitative terms. Nevertheless the concept of balanced risk is useful. It can be used to show that attempts to reduce low-level risks further are unwarranted compared with higher level risks that are accepted elsewhere. It can also be used to show how risk associated with the introduction of new technology (nuclear power, multiple-purpose dams, chemical industries, raw materials processing) compares with other and longer established risks (coal generation plans, dry-farming, or rain-fed agriculture in semi-arid or sub-humid regions, monoculture and plantation agriculture, and so on).

The concept of balanced risk however does not by itself provide an all-embracing guide to safety judgements. Even if a fully quantified balance of risks could be achieved (which it cannot), there remains the questions of benefits and consequences. It is difficult to consider alternatives merely in terms of risks, as their benefits rarely appear so equivalent that they can be excluded from assessment. The assessor is therefore almost always weighing the risks versus the benefits for each alternative.

2.4.3 Comparative Risks: Analysis of Other Risks

A common yardstick for measuring the significance or acceptability of a risk is to compare it with other risks. In this kind of analysis, the consequences are reduced to a common denominator — usually death — and the benefits are ignored or not specified. This is because the benefits cannot be strictly compared. The risks being compared are not alternatives in that they could each provide the same benefit.

Table 2.3 illustrates two such sets of comparisons in which risks are compared in terms of the probability of death per billion persons with one hour of exposure and one in one million probability of death. The risks vary from fatal snake bites, to coal mining, to being vaccinated or riding a motor bike. The benefits associated with each one of them are in no way comparable.

Table 2.3   Comparative Probabilities of Death for Different Activities

	Deaths per billion persons with one hour risk exposure
Being vaccinated or inoculated	1.3
Exposure to radiation in a two hour high altitude flight during solar flare	2.5
Living in area where snakes are present	3.8
Radiation exposure of world population in major nuclear war (areas away from conflict)	5.0
Railroad or bus travel (USA)	10.0
(Britain)	50.0
Child asleep in crib	140.00
Being struck by lightning	200.0
Coal mining (Br.)	400.0
Amateur boxing (Br.)	450.0
Climbing stairs	550.0
Coal mining (USA)	910.0
Hunting	950.0
Automobile travel	1200.0
Air travel	1450.0
Cigarette smoking	2600.0
Mountain climbing (USA)	2700.0
Boating (small boats)	3000.0
Motor scooter riding (Br.)	3000.0
Swimming	3650.0
Motor cycle riding (Canada)	4420.0
(USA)	6280.0
(Br.)	6600.0
Armed forces in Viet Nam	7935.0
Canoeing	10000.0
Motor cycle racing (Br.)	35000.0
Mountain climbing (Alpine)	40000.0
Professional boxing	70000.0
Being born	80000.0

One in a million risk of death from the following:

1½ cigarettes
50 miles by car
250 miles by air
1½ minutes rock climbing
6 minutes canoeing
20 minutes being a man aged 60
1 or 2 weeks' typical factory work

Sources: Stannard, 1969; *Insurance,* October 25, 1969; Pochin, 1974.

Table 2.4 Comparative Occupational Accident Rates

Federal Republic of Germany (1970)	
	d/M/y
Chemical industry	103 ± 11
Iron and metal industry	114 ± 6
Construction	274 ± 10
Mining	590 ± 47

France, National Statistics of Occupational Accidents (Statistiques Nationales d'Accidents du Travail, 1968-70)	
	d/M/y
Clothing industries	17
Textile industries	42
Metal workers	118
Chemical industries	169
Quarrying, etc.	365
Dockers (marine)	1020
Trawling, télépheriques, pleasure vessels, etc.	1636

United States fatal accident rates (d/M/y) (United States National Safety Council)

	1955	1958	1961	1964	1968	1971
All industries	240	220	210	210	190	180
Trade	120	90	90	80	70	70
Manufacture	120	120	110	100	90	100
Service	150	140	130	130	120	120
Government					130	130
Transport and public utilities	340	330	430	440	380	360
Agriculture	550	570	600	670	650	670
Construction	750	740	740	730	740	710
Mining and quarrying	1040	960	1080	1080	1170	1000

Source: Pochin, 1974.

These comparisons are actuarial ones, such as might be used by insurance companies to calculate rates for providing protection against them. When such a wide range of potential risk is included in the comparisons, the statistics are less useful and less valid for regulatory decisions — because no one decision maker has power to control so many different activities and the numbers may not be comparable — than for highlighting particular hazards, such as smoking, in order to make the public aware of them.

Within a more closely defined set of risks, however, such comparisons can be a useful basis for decision-making. Occupational risks are frequently compared in this way and the variability of the risks from year to year can also be included. Comparisons can also be made between countries for the risks of similar occupations (Table 2.4).

The range of risks involved can sometimes be attributed to the inherent riskiness of the activity — for example, rock climbing or race-horse riding, which both entail high risks of death or injury. It also reflects a more significant fact — that more effort is put into preventing loss of life in some occupations, and for some risks, than for others. In the United Kingdom, it has been shown that extraordinary effort is put into nuclear safety, even for the most unlikely accidents, compared to little effort to protect workers against accidents in agriculture (Sinclair, Marstrand and Newick, 1972).

Comparisons of the probability of death or injury between different occupations give a measure of the implicit value that is put on human life and injury by showing how much money is spent to prevent them. Thus where the risk level is very low, as in nuclear energy production or drug manufacture, the implicit value put on life is very high (over one million pounds sterling in UK, in 1972). In agriculture, it is low (only £10,000 per life in UK in 1972) and in trawling the value is zero or even negative (Table 2.5).

Table 2.5   Life Valuations for Different Occupations in the United Kingdom Derived from Risk Levels Set by Current Control Techniques

Occupation	Annual Risk of Death	Implicity Valuation of Life (£ sterling in 1972)
Trawling	1.4 in 1,000	negative value
Agriculture	2.0 in 10,000	£ 10,000
Steel handling	2.2 in 10,000	£ 230,000
Nuclear Energy		£ 1,000,000
Pharmaceuticals	2.0 in 100,000	£10,500,000

Source: Sinclair, Marstrand, and Newick, 1972.

Comparative data of this kind can therefore be used to redirect priorities to areas where better safety measures are needed and perhaps also where more stringent government surveillance is necessary. For example, in the UK, it is known that different government regulatory agencies also show a range of probability of death and injury for the occupations under their authority (Table 2.6).

Transportation accident rates are often put into a comparative framework, as within limits, different modes of transportation are substitutes for one another. One example for the UK is given in Table 2.7 in which the risk is first defined as the number of deaths per hundred million passenger-miles. However, this comparison masks the fact that, for aviation, landing and take-off represent times of considerably higher risk so that the accident rate is probably more related to the *number* of flights flown rather than to the distance covered, or to the time travelling. For 1974 in the UK, the risk in aviation per number of flights was 2.9 fatal accidents per million flights. This produces a quite different set of comparisons, if number of journeys are compared rather than passenger-miles which are shown in the lower part of Table 2.7.

The basis of comparison can therefore be crucial to the rankings of the risks being compared, and the statistical evidence presented can make planes seem safer or more dangerous than cars.

This is well illustrated in another area where comparative risk accounting has been relatively well quantified — that of energy production. Recent

Table 2.6 Death and Injury Rates per hour per Billion (10^9) Population at Risk for Occupations Regulated by Different Government Agencies in the UK (1966 data)

Inspecting Agency	Death Risk	Injury Risk
Offices, Shops and Railway Premises Inspectorate	1.85	1,200
Factory Inspectorate	38.95	16,500
Railway Establishments Inspectorate	144.00	16,000
Mines and Quarries Inspectorate	140.15	145,500
Agricultural Inspectorate	161.90	14,500

Source: Sinclair, Marstrand, and Newick, 1972.

Table 2.7 Number of Deaths by Various Means of Transport in the UK in 1974 Using Different Bases for Comparison

		Deaths/100 Million Passenger Miles
1.	Motor cycle	32.0
2.	Pedal cycle	13.3
3.	Motor Car Drivers	1.3
4.	Heavy Truck Drivers	0.8
5.	Motor Car Passengers	0.455
6.	Aviation (world wide schedule)	0.38
7.	Train Passengers (1973)	0.23
8.	Public Vehicle Drivers	0.2
9.	Public Vehicle Passengers	0.15

Passengers using:		Deaths/Million Passenger Journeys
1.	Aviation (475 miles av. journey)	1.8
2.	Trains (26 miles av. journey)	0.059
3.	Motor car (5 miles av. journey)	0.027

Source: Warren, 1977.

analyses have been conducted by government agencies in Canada (Inhaber, 1978) and the United Kingdom (Health and Safety Commission, 1978). They demonstrate some of the difficulties and possibilities in comparative risk assessment.

The Canadian study attempts to compare conventional sources of energy (coal, oil, natural gas and nuclear) with non-conventional ones (wind, solar, methanol and ocean thermal). It presents data on the occupational and public health risks of the total fuel cycles (see Chapter 3). Eight main sources of risk with the fuel cycles are considered:

(1) The production of fuel
(2) The production of other materials to build the hardware components required (e.g. nuclear reactor, solar cells)
(3) Manufacturing of components
(4) Construction of plant
(5) Operation and maintenance of plant
(6) Pollution and accident risks to public including catastrophic accidents
(7) Transportation
(8) Waste disposal

For each type of activity — mining, manufacturing, etc. — labour statistics were used to obtain the number of deaths, injuries and disease-related time lost per unit of time worked. Inhaber argues that energy generated by natural gas has the lowest overall risk associated with it, followed by nuclear energy and ocean thermal. According to his analysis, coal and oil generation of electricity have the highest risk and are 100 times more dangerous. However, the

Table 2.8 Deaths from Accidents Involved in Energy Production from Coal, Oil/Gas, and Nuclear power in UK

Energy source	Operation	Deaths/GWy from accidents
Coal	Extraction	1.4
	Transport	0.2
	Generation	0.2
	Total	1.8
Oil/Gas	Extraction	0.3
	Transport	insignificant
	Generation	none reported
	Total	0.3
Nuclear	Extraction	0.1
	Transport	insignificant
	Generation	0.15
	Total	0.25

Source: UK Health and Safety Commission, 1978.

Canadian study by Inhaber has received widespread criticism of its assumptions, methods and calculations; and it is apparent that such risk comparisons are extremely difficult, if not impossible, to carry out at present.

The British study (UK Health and Safety Commission, 1978) also tried to compare the total social costs of different means of energy production but because of the uncertainties, decided to exclude health effects from pollution. It therefore restricted itself to accidents and even excluded severe, rare accidents (Table 2.8).

Some of the hesitations expressed in the British study illustrate the difficulties and simplifying assumptions involved in this kind of exercise.

(1) Data for non-conventional technologies are too poor to include them.
(2) Data for health effects (e.g. through air pollution) too unreliable to include them.
(3) Severe, rare accidents excluded because assumed probabilities are too low to affect results.
(4) Accident data for uranium mining in USA assumed similar to Australia, Canada and South Africa (where British uranium comes from).
(5) Effect of low-level emissions of radiation included but *not* effects of low level emissions of heavy metals from coal, gas and oil combustion.
(6) Dose commitment for future potential health damage of radiation calculated but not for other fuels.
(7) Deaths and injuries not added together (Canadian study values one death at 6,000 man-days lost).

The British report concludes that:

'If sensible comparisons are to be made between the environmental consequences of nuclear energy and those from other fuel sources, the methodology has to be resolved'.

Comparative risk assessment is still in its infancy and all such data should be carefully scrutinized for simplifying assumptions, non-comparability of data, and the validity of the units in which the risk or loss is expressed. Furthermore, data themselves are often critically lacking.

2.4.4 Benefit-Risk: Consideration of Benefits

A fourth way to compare risks is to compare them with the benefits they bring. By this argument greater risks can be accepted where there are greater benefits. An extreme example is the utilization in medical practice of high risk surgery or high risk chemotherapy or radiation treatment to prolong the life of terminally ill patients. The impending threat of death allows for the rational choice of more extreme risks in treatment than would otherwise be acceptable. Another common basis for comparison is to set the benefits of higher wages against the higher risks of some occupations (Figure 2.6).

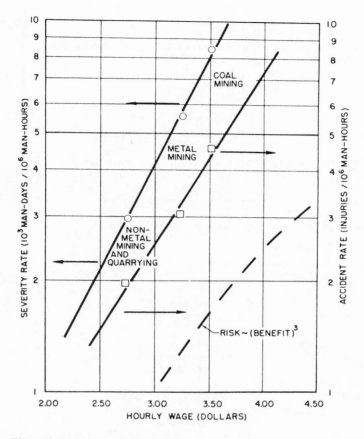

Figure 2.6 Mining accident rates vs. monetary reward (risk-benefit)

Source: Starr, 1972. (Reproduced from *Perspectives on Benefit-Risk Decision Making* (1972) with the permission of the National Academy of Sciences, Washington, D.C.)

A benefit-risk argument is also commonly employed in supporting major engineering works for development, in nuclear energy programmes, in pesticide applications and in other instances. It is claimed that the ecological risks attendant upon major dam construction are acceptable because the benefits from an expanded acreage of irrigated land will be large. Similarly, others would argue that the risks of major accident or massive release of fission products from nuclear installations are acceptable because of the benefits of increased power supply. Again the risks to ecosystems and perhaps to human health, from pesticide use are acceptable to many governments and individuals because the loss of food production that would accompany failure to control pests is a greater and more serious risk.

Table 2.9   Risk Assessment Inventory for DDT made by US Environmental Protection Agency 1975

Benefits	Risks	Data Exper.	Source Environ.
Improved yields in $^1/_6$ US cotton acreage	Bioaccumulates in living organisms up food chains to man		X
Saved \$1.00-\$6.00 production costs per treated acre of cotton (US total cost \$54 M p.a.) Translates into increased consumer costs of 2.2 cents per capita per year	Is a persistent substance and can be widely distributed in air, soil, and water far from points of application		X
	Is everywhere in the environment		X
Important available substitute (methyl parathion) principal pesticide cause of human poisonings	Is common in human food		X
	Is stored in human tissue		
Improved yields on variety of vegetable crops	Is a *potential* human carcinogen (experimental data on mice)	X	
Only effective pesticide for pea leaf weevil	Decreases photosynthesis by photoplankton and adversely affects growth rate	X	
Only effective pesticide against two forest pests (Tussock and Gypsy Moths)	Has lethal and sublethal effects on useful aquatic freshwater invertebrates	X	X
Exterminator for bats and mice used by military	Can kill most fish species at very low concentrations	X	
Disease vector control (malaria) for public health (held in reserve for emergencies in US)	Has been responsible for fish kills		X

The principal, and often overriding difficulty in a risk-benefit approach is that the two sides of the equation are almost impossible to compare quantitatively or comprehensively. Furthermore, the benefits are usually more quantified and demonstrable than are the risks. The balance in a risk-benefit analysis is thus often tilted towards the benefits, especially at the beginning. Over time, the perception of risks and benefits changes and as new data on risks become available (through monitoring, testing and general experience) the balance may become tilted towards emphasizing the risks.

Risk-benefit analyses for major technologies such as a large industry or irrigation scheme, or for new products such as drugs or pesticides, can suffer from major deficiencies of data and high uncertainty. For new chemicals, the

Benefits	Risks	Data Exper.	Source Environ.
	Affects reproductive process fish	X	X
	Has adverse physiological and behavioral effects on fish	X	
	Causes thinning of eggshells and interferes with reproduction of birds	X	X
	High concentrations in birds of prey through bioaccumulation threatening species survival		X
	Many pests now resistant to DDT	X	X
	Alternative pesticides are effective at acceptable cost for cotton and most vegetables	X	X
	Alternative exterminators are effective and available	X	X

Decisions:

1. Ban DDT as pesticide in USA except
   - a) Allow temporary use against pea leaf weavil in 1974
   - b) Allow emergency use against Tussock Moth in 1974
2. Ban DDT as exterminator for bats and mice
3. Keep DDT in reserve as disease vector control for public health
4. Allow export of DDT

Source: US Environmental Protection Agency, 1975.

risks are largely unknown at the outset while the benefits may be loudly acclaimed. Increasingly, the marketing of new products is preceded by careful testing. However with carcinogenic hazards, the time delay between the benefits (employment or product) and the risks (cancer) may be years, or decades, or even generations. The benefits and risks may fall on different groups of people — occupational risks are concentrated on workers while benefits may be widely spread among the public. Other risks, such as air pollution from industry, fall generally on the public but the benefits may be concentrated in the private sector.

There are many problems in drawing up a balance sheet for the risks and benefits associated with a particular hazard. Some of the necessary decisions involved have been discussed earlier in this chapter. Producing a risk-benefit inventory can, however, provide valuable input to the decision-making process. It provides a summary overview of the issues, and can make explicit what has been included, and what omitted, from consideration. Take, for example, the accounting risks and benefits for DDT that led to the decision by the US Environmental Protection Agency to prohibit the use of DDT as a pesticide in the USA (Table 2.9). Here the number of risks would appear to outweigh the benefits, but the benefits are better quantified. The lack of effective alternatives for certain pests proved to be a major factor in the exemptions permitted in the overall ban on its use as a pesticide, and the small environmental impact of public health uses in the US, led to its continued use (in reserve) as a disease vector control.

Risk benefit comparisons can also include the risk of *not* taking or doing something. This has entered into the present debate about the risks of saccharin, a low calorie sweetner which has been linked experimentally to bladder cancer, and which has been suspended from use in soft drinks in Canada and USA. The suspension was based on a risk assessment that if the US population were to drink one diet soft drink each day throughout their lives, there would be an extra 1,200 bladder cancers over the whole US population per year. This leads to a calculation that on average one diet drink would reduce life expectancy by nine seconds. The benefits of diet soft drinks are that they help to control obesity. Cohen argues that a 45 year old male weighing 170 pounds instead of an optimal 150 pounds has a reduced life expectancy of four years. Thus a diet drink has only to reduce his calorific intake by 1 calorie per day to make the benefits of saccharin outweigh the risks. Furthermore, diet soft drinks have about 100 less calories per drink so that *not* using saccharin in a soft drink increases personal risk by a factor of 100 (Cohen, 1978). The initial risk assessment concentrated only on the *risks* of saccharin and ignored the benefits. On the other hand, the risks from saccharin *and* obesity can both be reduced by drinking water!

Risk-benefit analysis can sometimes be highly quantified but more often its value lies in its comprehensive approach than in its statistics. Even where they cannot be quantified, risks and benefits should be considered together within a single accounting system and in some circumstances formal techniques of risk-benefit analysis can be applied.

CHAPTER 3

Identifying and Estimating Risks

3.1   THE SELECTION OF TECHNIQUES

The estimation of risk is concerned with collection information on:

(1) The nature and extent of the *source;*
(2) The chain of events, *pathways* and processes that connect the cause to the effects; and
(3) The relationship between the characteristics of the impact (dose) and the types of response *(effects).*

Although response and effects are often equated or used synonymously, the World Health Organization makes a distinction between the two (WHO, 1972). An *effect* is a 'biological reaction itself, while a *response* refers to the relative number of exposed people who react with a specific effect' (Friberg, 1976). This concept is illustrated in Figure 3.1. According to some scientists,

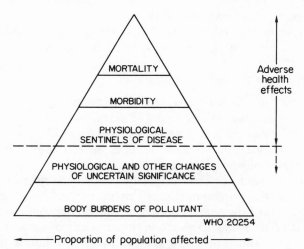

Based on a diagram in United States Congress Document No. 92 241, 1972

Source: Munn, Phillips and Sanderson, 1977

Figure 3.1 Schematic representation of the WHO concepts of effect and response (WHO, 1972)

the WHO distinction is useful and should be extended to other kinds of receptors such as vegetation (Munn, Phillips and Sanderson, 1977).

In practice, risk assessment often begins by looking at one part of the problem, usually the source or the effect, rather than considering the system as a whole at the outset. This is a pragmatic response to the different ways in which risks are discovered. Some risks are identified initially as a known or suspected (such as a carcinogen) source, and the problem is to discover where and how the effects are distributed. Other risks emerge because their effects become apparent (such as a localized high incidence of a disease) and the priority risk assessment task is to determine the cause or source of the problem.

Once a risk is suspected, it is important to bring together as much available information as is possible before designing the ways in which additional data are to be collected. The estimation of risk is an expensive process and initial errors in the design of monitoring or screening programmes are costly in money, delays and inadequate information.

Some of the initial questions that should be asked before any assessment methods are selected are:

(1) Is the problem an analysis starting from a suspected (or known) *cause* or *effect?*
(2) How is the problem distributed geographically? (highly localized, regional, national, associated with urban areas, in lowlands, etc?)
(3) How much time available is there to assess the risk? (fast or slow causes, acute or chronic effects?)
(4) Where do the effects lie? (within which demographic populations or which parts of the environment?)
(5) Is the target (receptor) normal or particularly susceptible? (healthy adults or sensitive groups; unstressed ecosystem or already highly polluted?)
(6) Are the effects reversible or irreversible?

By specifying the problems in terms of questions like these, the decision maker can arrive at some understanding of the magnitude of the task, where to start, and how much time is available. His answers will help him to select the appropriate monitoring or testing techniques as well as provide him with a rudimentary model of what is going on.

It is important also to be able to specify whether the risks arise principally out of a *technological system;* or through *environmental processes,* or through *human biology and behaviour.* This is because the appropriate measurement techniques differ according to where in the sequence of cause and effect you wish to measure. Many risks, of course, involve technical processes and human behaviour as well as environmental processes.

For example, lead pollution from an industrial lead smelter may be caused initially from poor system design and equipment malfunction and here techniques appropriate to technical systems would be appropriate (e.g. fault tree analysis, components testing). Once the lead leaves the chimney stack, it is subject to environmental processes of transportation and diffusion through the air, settling to the ground, being taken up by plants and animals and movement through the soil and water. Quite different models and measurement techniques from those applicable within the smelter, would be used. How much of the lead concentrated in vegetables or drinking water eventually reaches individual people depends on another set of human biology and behavioural variables, such as how much contaminated food they consume. Thus models and techniques to measure human behaviour are relevant at this point in the sequence.

The selection of techniques thus requires initial decisions to be made about:

(1) The main methods — monitoring, testing, and modelling.
(2) The focus of concern — technical, environmental, or human behaviour subsystems.

Table 3.1 Examples of modelling, monitoring and testing techniques for estimating environmental risks

| Approaches | Environmental Risk System | | |
	Technical	Environment	Human Biology and Behaviour

Modelling	Fuel cycles
	Event tree analysis
	Fault tree analysis
		Ecosystems models
		Physical transport
		  models
		Food chain models
		Tectonic models
		Hydrologic and
		  atmospheric
		  models
			Genetic models
			Demographic models
			Metabolic pathways
			Epidemiological
			  models
			Behavioural models

Monitoring
 	
and
 	
Surveillance
	System reliability
	Pollutant emissions
	Component quality
	Design quality
		Weather watch
		Seismic movements
		Air quality
		Forest and crop
		  surveys
			Mortality statistics
			Notifiable diseases
			Clinical records
			Drug use reporting
			Epidemiological
			  surveys

Screening
 	
and
 	
Testing
	Materials testing
	Product quality
	Destructive tests for
	  failure
		Screening tests for
		  persistence
		  transformation
		  toxicity
		Recovery rates
			Mutagenicity (Ames
			  test)
			Acute toxicity
			LD_{50} screening test
			Response mechan
			  isms
			Physical check-ups

Table 3.1 gives examples of available techniques according to the main approaches and subsystems. The main approaches discussed here are environmental monitoring and health surveillance, testing and screening, and modelling.

3.2   ENVIRONMENTAL MONITORING AND HEALTH SURVEILLANCE

Environmental monitoring is defined as 'the process of repetitive observing, for defined purposes, of one or more elements or indicators of the environment according to prearranged schedules in space and time, and using comparable methodologies for environmental sensing and data collection' (Munn, 1973).

Environmental monitoring is undertaken to determine space and time patterns of environmental elements or indicators, and to estimate their variabilities, for the purpose of:

(1) Understanding environmental processes:
(2) Providing early warnings of environmental threats (natural as well as man-induced);
(3) Assisting in the optimization of the use of renewable and non-renewable resources;
(4) Assisting in the regulatory process, providing data that may be used in the courts or elsewhere to demonstrate cases of non-compliance with environmental standards.

In the case of (2) above, the early warnings may be given by simple extrapolations of upward trends, or preferably from predictions obtained from models that have been already validated with independent sets of data.

Within individual countries, there are many kinds of monitoring programmes which are often uncoordinated with each other and with similar programmes in adjacent countries. In particular, the instrument and sampling techniques may be so disparate the inter-comparisons may be impossible. Yet for environmental risk assessment, the need for reliable data is essential.

There are several multi-national monitoring systems that have been established for particular reasons. Examples include the OECD and EEC long-range transport of air pollution programmes and the Lake Erie International Field Year on the Great Lakes. In most cases, these monitoring systems have been established to determine the existence and/or extent of an environmental problem that has been perceived in general terms only.

Some components of the national and regional programmes, are connected globally through the United Nations Environment Programme in the Global Environmental Monitoring System (GEMS). This is a coordinated programme for gathering data to be used in environmental management (including early warning systems) rather than for enforcement of existing environmental standards. The data will also be valuable in some cases in the development of legislation or international conventions leading to the establishment of environmental controls.

Monitoring and sampling programmes are also used to detect the presence of harmful substances. For example, the discovery of small quantities of carcinogens in the drinking water of several cities in the United States was made during a routine sampling programme to determine water quality.

Research into the source of these contaminants has led to a re-evaluation of current drinking water disinfection practices in the United States, as the source of these contaminants appears to be a by-product of the disinfection practices.

Monitoring of administrative programmes such as the required registration of chemicals, drugs, pesticides and industrial undertakings can also be important. For example, the information submitted to obtain a permit or to register a toxic material can help identify the nature and amount of specific pollutants entering the environment.

Health surveillance is the collation and interpretation of data collected from monitoring programmes and from any other available sources, with a view to detecting changes in the health status of populations (WHO, 1972). Internationally, health surveillance has been most successful in recording the principal causes of death and notifiable diseases such as malaria or smallpox. There is a need however to maintain surveillance programmes that will alert governments and scientists to environmental hazards when symptoms first appear rather than when the patients have died.

One important use of surveillance is for the effects of drugs. In the UK a system of reporting about drug use and effects from doctors to a central agency is a principal means of monitoring and evaluating drugs. The incidence of cancer, and accidents (home, industrial and transportation) are also regularly monitored in many countries.

Monitoring human populations for genetic mutations is of concern in some industrial countries where the chemical environment is rapidly changing and becoming more complex with increasing exposure to chemical mutagens. The counting of spontaneous abortions has been recommended as a simple and practical method for monitoring mutation rates. The spontaneous incidence of abortions is about 1 per cent of all births.

For all these health surveillance programmes, the lack of baseline data on regional and national incidences of health effects, the need for skilled manpower, effective reporting systems and centralized data bank facilities, makes progress slow in all countries. Where detailed health data are already centralized, for example, for financial purposes in national health schemes, they can also be used to help establish the effects of the environment on health.

3.3   TESTING AND SCREENING

If based on accepted, standardized procedures, testing can identify sources of risk (such as defective products or dangerous chemicals) or harmful effects (such as acute toxicity and mutagenicity). Tests can also help to quantify environmental processes such as the persistence of harmful substances in the environment. Testing is often a carefully controlled laboratory procedure. The advantage of field or in-situ testing is that it provides more realistic conditions.

Screening usually involves multiple tests or multiple candidates for evaluation which are 'sifted' by the test procedures into different categories. In numerous instances, tests are carried out for product or drug or food-additive safety. This is often done with animal test populations, and from this, inference is made of the degree of risk to human populations. In the United States, for example, the Food and Drug Administration screens some 110 petitions for new food additives annually, and the Environmental Protection Agency has been charged under the recent Toxic Substances Control Act (1976) with the formidable task of screening the 30,000 chemical substances now in substantial use and the several hundred new ones being added each year.

It is, in fact, impossible to keep pace with all the products or chemical substances that need to be screened and tested. The United States has fallen behind in this operation and many other countries are even further behind. The cost of tests is another consideration to be added to those of the time they take and the skilled manpower and equipment needed. For example, it was estimated that the basic tests for registering a pesticide in the USA in 1970 cost

Sub system

Table 3.2 Basic Tests Required for Pesticides Registration in the USA (from Blodgett, 1974.

	Tests	Date established	Cost (estimated dollars)	Time implication
1.	Chemical and physical properties (Such as solubility, vapor pressure, flash point)	1947	5,000-15,000	
2.	Degradation studies			
	Persistence (soil)	1965		6-24 months
	Persistence (water)			Less than 1
	and sediment)	1970		year
	Photochemical	1970		2-6 months
3.	Mobility studies	1970		Less than 6 months
	Runoff			
	Leaching	1970		Less than 3 months
4.	Residue studies			
	Fish	1970		2-6 months
	Birds	1970		2-6 months
	Mammals	1970		2-6 months
	Lower trophic levels of food chains	1972		6-9 months
5.	Microbiological studies	1970		Less than 3 months

A rough estimate of these requirements in their entirety would range between $100,000 and $250,000.

It should be noted that much of the data generated by these tests is utilized in studies of human, fish, and wildlife safety.

between $100,000 and $250,000 and required 1-2 years (Table 3.2). In addition, tests to measure the tolerances of other plants and animals to pesticides cost between $5,000 and $160,000 per test and take between two and 28 months to complete (Table 3.3).

These high cost and time demands of tests have encouraged the development of more rapid and less expensive tests. For example, the current approach in testing effluent is to break-down the effluent into its chemical components and screen each for toxicity, persistence and breakdown or transformation properties. This bank of tests is necessary because it is not sufficient to know just the chemical properties at the point of discharge. Certain chemicals will breakdown or transform *after* they leave the point of discharge into substances which may be more or less harmful than the original chemical. For example, the pesticide myrex, when transmitted through a water environment will break down into kepone, a highly toxic and persistent substance. Thus tests must be run to determine if particular contaminants will break down and if so, into what. Tests must also be run to see if the substance is likely to be captured in sediments and there build-up, or bioaccumulate in various animal and plant forms.

Table 3.3 Tests for Tolerance of Pesticides in USA (from Blodgett, 1974. Reproduced by permission of the MIT Press)

	Tests	Date established	Cost (estimated dollars)	Time implication (months)
1.	Toxicology			
	Acute (rat and non-rodent)	1954	5,000	1
	Subacute (rat and dog)	1954	50,000	6
	Chronic, 2 year (rat and dog)	1954	160,000	28
2.	Reproduction (rat)	1960	35,000	20
3.	Teratogenesis	1970	10,000	2
4.	Mutagenesis	1972	10,000	2
5.	Metabolism			
	Plant	1954	50,000	6
	Animal	Before 1960	25,000	3
6.	Analytical			
	Methodology	1954	100,000	4-6
	Crops, Meat, Milk, Poultry,			
	Eggs	1965 (Poultry)		
7.	Field Residue Data		100,000	12
	Drop, Feed, Meat,			
	Milk, Poultry,		100,000	6
	Eggs	Before 1960 1965 (Poultry)		

Theodore Lownik Library
Illinois Benedictine College
Lisle, Illinois 60532

The ability to test for the various parameters is uneven. The chemical/ physical parameters can be measured rather well. Measuring the biological parameters is much more difficult. Fairly good progress has been made in developing tests for fresh water, and a beginning has been made in the marine area. However, considerable research remains to be done in the developing of suitable biological screening techniques.

In order to streamline the process, rapid screening tests are being developed to screen out those components most likely to be a problem. Those chemicals identified as bad actors as a result of the rapid screening tests, are then put through more detailed tests to develop more refined data on impact levels. These and related procedures are more fully discussed in another SCOPE report (Butler, 1978).

Continuing on the current chemical specific basis, however, is extremely cumbersome because of the sheer number of substances to be considered. Thus, research is also being conducted to see if techniques can be developed to evaluate the effluent as a totality, to make the problem more manageable while still maintaining reasonable safety factors. To make this approach feasible a similar bank of tests for toxicity, transformation and persistence would have to be developed to handle the *total effluent*. This will reduce the regulation task considerably if effluent does not have to be broken down into its components to show that it is harmful.

3.4 MODELLING

All the risks that are of concern in this report arise as part of a process. To understand the risk and to manage it effectively models of the total process are developed. Such models have, or should have, both environmental and social components, because the degree of risk depends both upon what happens in the environment and what social processes are at work.

Highly successful models have been developed for some environmental processes, especially in such areas as meteorology (atmospheric diffusion models), hydrology (run-off and flood forecasting models), and plant growth (biomass production models). Capability to model the social systems and how they interact with environment is much less well developed. Thus the more important the social component or human behavioural component becomes in an environmental risk, the less likely we are to be able to model the total process satisfactorily.

Models are developed in every branch of science that bears upon environmental risk questions. The many possible models may be grouped, however, into those that are primarily technological in orientation, those that are primarily concerned to describe environmental processes and those that are concerned with human systems whether social, biological, behavioural and so on.

The focus in this report is on the environmental models, but the environmental models are often connected at one point or more to

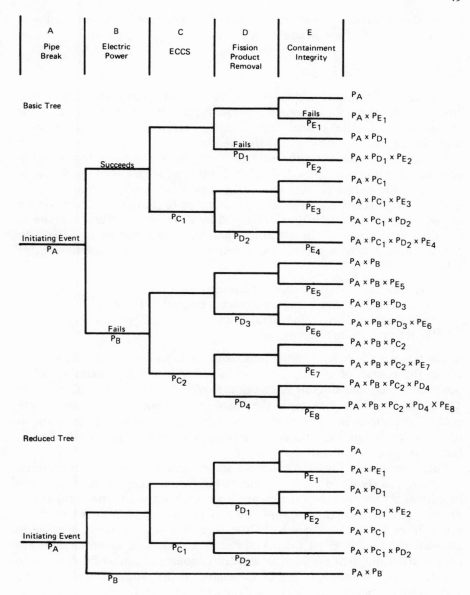

Note - Since the probability of failure, P, is generally less than 0.1, the probability of success (1-P) is always close to 1. Thus, the probability associated with the upper (success) branches in the tree is assumed to be 1.

Figure 3.2 Simplified Event Trees for a Large LOCA

Source: US AEC, p.89 (Reproduced by permission of the U.S. Atomic Energy Commission.)

technological and social models. It is important therefore to recognize some of the characteristics of the different kinds.

3.4.1 Technological Models

In the case of technological models all the links and pathways of the process are known in principle because the whole process has been designed by the human brain, but the probabilities of different failures, particularly those involving a series of malfunctions, are unknown.

A method of estimating the probability of failure in large technological systems is event tree analysis. A simplified event tree for a large LOCA (loss of coolant accident) at a nuclear generating station is shown in Figure 3.2. An initiating event is postulated — in this case a pipe break in the primary system of the reactor. The tree is then developed by determining, from an intimate working knowledge of the reactor, what other systems might affect the subsequent course of events. In the example given, the main questions being asked are

'Will the station's own supply of electric power fail?'
'Will the ECCS (emergency core cooling system) fail?'
'Will fission product removal fail?' and
'Will containment integrity fail?'

The systems are ordered in the time sequence in which they are expected to affect the course of events. For the initiating event a probability of its occurrence is estimated. How often can such pipe breaks be expected to occur? This is obtained by experience in the use of pipes in non-nuclear systems. For each of the succeeding events the probability of the system to perform its function (probability of success) is estimated as well as the probability of failure to perform its function.

The upper half of Figure 3.2 shows a set of theoretical paths or event trees. In practice, as shown in the lower part of the figure, many of these paths can be eliminated because they represent illogical sequences. If the event tree is properly constructed (the example here is greatly simplified) the series of events in each accident chain is defined so that it is in principle possible to calculate the consequences for that series. The event tree approach can thus provide a definition of the possible accident sequences from which radioactive releases to the environment can be calculated. If the failure probabilities are known, the probability of each release can be calculated (US AEC, 1974).

3.4.2 Links to Environmental and Behavioural Models

At this point the model of a complex technological system connects with the atmospheric diffusion model. If radioactive fission products are released how far and how fast will they spread? This leads in turn to questions about effect on human populations for which models of the response of the human biological system to radioactivity are needed as well as models of human behaviour in the face of warnings about radioactivity in a power station accident. These behavioural elements may have substantial effect on the amount of damage to health that is caused, but are more difficult to model in a reliable fashion.

An important source of difficulty in modelling the human behavioural component is that it is self-aware. Accident sequences in nuclear power plants, and the process of biomagnification of toxic substances in food chains all proceed in an automatic fashion which can be modelled in probabilistic terms. The quality of self-awareness in human systems enables the behaviour and functioning of the system to change in response to how people think the process is working or should work. In other words, we now need to know what perceptions of the system are held by the participants in it, since that will change the outcome.

3.4.3 A Perception-Behaviour Model

A model for environmental perception and behaviour is shown in Figure 3.3. It represents one way of organizing the components of a general model (Whyte, 1977). The variables are arranged according to:

(1) The distance from a decision point at the man-environment interface, and
(2) The scale at which decisions are taken from individual or household level to the organization or government level where decisions are taken by a few on behalf of many.

Thus as one moves from right to left across the diagram the variables more directly impinge upon the output variables for a specific intervention (but they may not necessarily be more influential). And as one moves from the bottom to the top of the figure the variables become more relevant to collective rather than individual decision-making, though they are not exclusive to either. Thus a progression can be traced from individual and group characteristics through intervening variables such as values and personality, to decision and choices affecting the environment.

Linking the individual and social variables are four interdependent processes which together act as the main organizing force in the system. These are the 'perception processes' which link all the components. In this model they are considered as four process elements on the pragmatic grounds of what are measurably different components of perception at the field level. Thus, *categorization* and *judgement* are grouped together in the model because they are often measured together although conceptually they are different parts of the perception process.

The other three major divisions of perception used here are: sensory perception (e.g. sight, smell); attitudes; and communication and information flow. In the field, these processes (either separately or together) can be

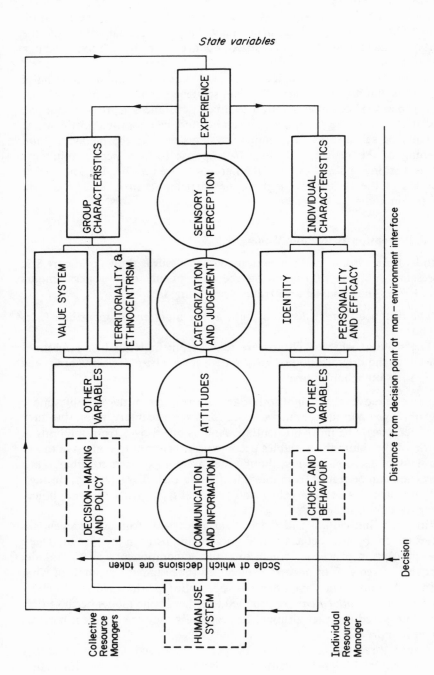

Figure 3.3 Simplified model of environmental perception
Source: Whyte, 1977

investigated as links between any sub-set of variables relevant to the study.

Figure 3.3 is a simple heuristic device to help organize the research planning task. It is not a model of how the system actually functions and it cannot serve as a substitute for specific hypothesis development and conceptual modelling within each research project. No constraints are intended on the boundaries of a specific study or on the operations and connections between variables. Specific environmental risk problems can therefore be examined in terms of their own 'critical paths' through the model.

3.5   ENVIRONMENTAL MODELS

A wide range of environmental models have been developed. The main purpose of environmental models in this context is to study the variables and linking processes between sources or causes of environmental risks and their consequences or effects. Much has been written elsewhere about the development of such models. This section illustrates environmental modelling by three types of example:

(1) Non-quantitative models and probabilistic models,
(2) Global models of biogeochemical cycles, and
(3) Models of persistent substances in food chains.

3.5.1 Non-quantitative Models and Probabilistic Models

The power and value of developing comprehensive models to examine events lies in their ability to portray in an understandable form complex systems and interrelationships. Their weakness is in the simplifying assumptions that are used in some or all of the component parts (Holdgate and White, 1977). Often these simplifying assumptions are made when processes (physical, biological or social) are represented in the form of equations. There is thus good reason to develop as a first step a non-quantitative model which maps out all the possible sources and pathways whether or not they can be quantified. Numerical representations of parts of the system can be developed in various ways as a second stage (including graphical mapping, empirically fitting algebraic equations to represent relationships between phenomena, and the use of high speed computers to solve complex systems of non-linear partial differential equations or stochastic relations representing such phenomena).

Figure 3.4 shows a non-quantitative model of the pathways by which DDT can reach the human body in a poor agricultural area. The pathways include contamination of DDT containers reused for storage of food and water and contamination of food when it is sold after it has been stored with DDT in village shops. Study of the behaviour of people using DDT or even those living in villages where it is used are as necessary inputs to the development of this kind of model (however unquantified) as is knowledge of the natural environmental and food chain processes. Indeed, comprehensiveness is a main objective of these models to ensure that all possible pathways by which risks

54

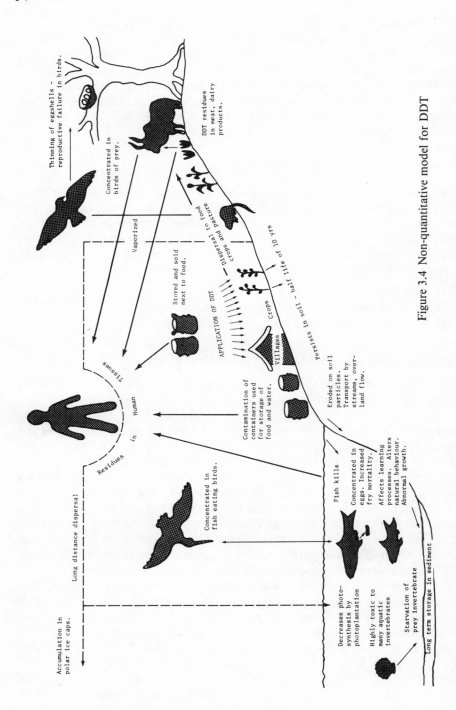

Figure 3.4 Non-quantitative model for DDT

are transmitted are included in the first picture so that the decision-maker is provided with an overall view of the magnitude and complexity of the problem he has to deal with. Later, simplifying assumptions will be introduced to enable processes to be quantified and parts of the model may even have to be omitted.

In many physical and a few biological applications, quantitative models are available that permit estimation of the *frequency distribution* of hourly, daily or monthly values of some environmental state or condition. For example, the frequency distribution of hourly concentrations of sulphur dioxide at a point in the vicinity of a chimney may be estimated, even though prediction for a *specific hour* could be wildly in error. From such calculation, the probability of any given concentration being exceeded can be specified.

3.5.2 Global Models of Biogeochemical Cycles

Some of the largest environmental modelling exercises now being attempted are in the area of biogeochemical cycles of nitrogen, phosphorus, sulphur and carbon. Understanding the volume of flows and the various pathways is essential to knowing how the cycles have been or might be changed by human action and what management actions might be possible or required. For example, the model of the global nitrogen cycle (Figure 3.5) underlies estimates of present and future possible contributions of N_2O to the depletion of the ozone layer. Similarly the development of knowledge of the sulphur cycle (see Figure 3.6) underlies the acid rain problem described in more detail in Chapter 6.

When a particular environmental problem arises, it may be necessary to combine parts of existing models into one that serves the purpose in question. For example the ozone layer depletion problem requires an understanding of atmospheric diffusion (how fast will chlorofluoromethanes and nitrous oxide diffuse up to the stratosphere?); it also requires a chemical interaction model that states the rate of interaction with ozone: thirdly, there is need to know the relation of ozone depletion to the amount of increase in ultraviolet-B radiation; finally a biological model is needed of the effect of increases in ultraviolet-B upon target organisms at the surface. Because of the uncertainty of estimates, models of this sort are continually being revised and updated as more observations become available.

3.5.3 Persistent Substances in Food Chains

Within the grand design of global cycles it is often necessary to examine small components to trace the entry and pathway of toxic substances. Food chains are a good example of this type of modelling.

In any ecosystem the energy captured by green plants becomes available in a stepwise fashion to animals and microorganisms by flowing through a food chain. There is an inevitable loss of available energy at each of the steps in the chain (Figure 3.7).

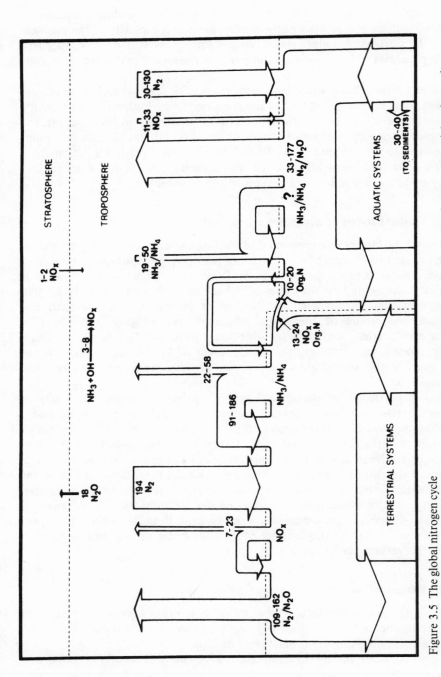

Figure 3.5 The global nitrogen cycle
The rates are given as Tg N yr^{-1}. The flows of N_2/N_2O are residuals obtained when balancing the terrestrial and aquatic systems (Soderlund and Svensson, 1976)

Source: Holdgate and White, 1977, p.40. (Reproduced by permission of SCOPE)

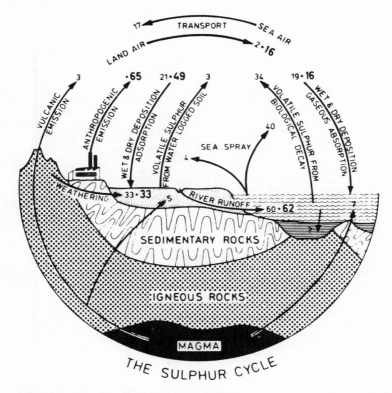

Figure 3.6 The first global sulphur cycle
Based on a preindustrial balance of the soil compartment. Fluxes are given
in Tg (millions of tonnes) of sulphur per year. The man-made parts of the
fluxes are indicated by plus signs (Hallberg, 1976)

Source: Holdgate and White, 1977, p.47. (Reproduced by permission of
SCOPE)

Two consequences of the ecological principle have significance in assessing
the risk to the environment of 'abnormal' levels of substances, and of
'abnormal' substances resulting from man's activities:

(1) Organisms at one level may be killed, breaking one link of the chain and
    thus disrupting the whole system.
(2) A substance which is persistent (not metabolised or biodegraded) may
    reach increasingly high concentrations as it is passed through the various
    levels of the chain. This process is referred to as biomagnification.

Persistent substances which have received attention recently include:

(1) Chlorinated hydrocarbon insecticides, e.g. DDT, dieldrin,
(2) Polychlorinated biphenyls (PCBs),
(3) Heavy metals, e.g. mercury, lead,
(4) Radioactive materials, e.g. Strontium-90, Iodine 131.

58

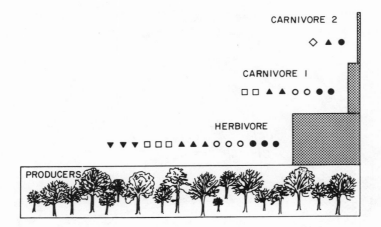

Figure 3.7 An intact natural ecosystem (e.g. a forest) where 10-20% of the energy at each level is passed to the next level. (Modified from Ehrlich *et al*, 1977)

Source: From *Ecoscience: Population, Resources, Environment* by Paul R. Ehrlich, Anne H. Ehrlich and John P. Holdren, W.H. Freeman and Company. Copyright © 1977

The first three are discussed briefly here as examples of food chain models that have been developed.

DDT With the information that DDT could be found in carnivorous birds at concentrations a million times greater than those in the abiotic environment (e.g. Woodwell, Craig and Johnson, 1977), it was clear that this particular compound did obey the laws which could have been predicted at the time of its first extensive use in 1948. The relevant properties of this man-made compound in this context are:

(1) It is chemically stable, very few organisms can degrade it. Its half-life (time for 50% to be lost from the original site of application) has been estimated at 10-15 years.
(2) It is readily dispersed in water and in air in the form of suspension and by codistillation with water into air.
(3) It is barely soluble in water, but soluble in lipids, which are components of all biological systems. Thus it will preferentially associate with living over non-living material.
(4) It has become ubiquitous.
(5) In addition to being bioaccumulated by individual groups of organisms, it is biomagnified in food chains. The aquatic food chain shown in Figure 3.8 exemplifies this.

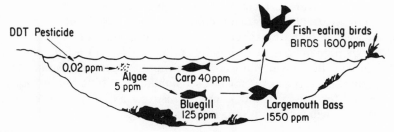

Buildup of chemicals along a representative food chain.

Figure 3.8 DDT in Clear Lake California (Modified from Treshow, 1976)

Source: From *The Human Environment* by M. Treshow. Copyright ©
1976 McGraw-Hill. Used with permission of McGraw-Hill Book Company

(6) It is a current constituent of the animal and human body (Wasserman,
Tomatis and Wasserman, 1975).

(7) It is transferred from mother to foetus through the placenta and to the
infant through breast milk.

(8) It activates or inactivates liver microsamal enzymes and thus interferes
with normal metabolic pathways. This fact explains the majority of its
biological effects.

(9) It is toxic not only to target organisms (the insect pests) but to a wide
spectrum of living organisms.

(10) Some insects can become resistant to the toxic effect so that other com-
pounds of higher toxicity are required to control the pests.

DDT was deliberately introduced for controlling insects and insect borne
disease (O'Brien, 1967) and for future insect control so here the message is
clear: there is an element of choice. The other persistent substances to be
discussed here have been released as a result of man's activities, but usually as
waste products.

Polychlorinated biphenyls (PCBs) PCBs share many of the properties of
DDT. They are chemically inert, extremely stable, have low solubility in water,
a high affinity for lipids and accumulate in living organisms from which they
are excreted only very slowly. They have become widely distributed through
the biosphere e.g. in the blubber of seals and porpoises on the East Coast of
Scotland and Eastern Canada. Thus they are prime candidates for food chain
accumulation and biomagnification and are now, like DDT, found in animal
and human tissues around the world. Well documented effects of PCBs on
human health include the rice-oil disaster in Japan (Yusho disease) which
affected more than 1,000 people who consumed food contaminated by PCBs
which had leaked into cooking oil from a heat exchanger (Kuratsune *et al.,*
1972). PCBs are also carcinogenic and are known to change the activity of the
liver and to interfere with normal metabolic pathways.

Sources include transformers, capacitors, plasticisers, solvents, adhesives, sealants, coatings for lumer and concrete, tires and brake linings. The compounds get into the biosphere by atmospheric emissions, dumping into water, and by human contact with paints and plastics. In Canada and the US they are not prohibited in food containers and processing plants. The US Environmental Protection Agency has set standards of 0.01 ppb in water and 0.5-1.0 mg/m^3 for air. Even with complete restriction on further use however, existing residues will persist for long periods of time.

Mercury While mercury has long been recognized as an extremely toxic substance (ReVelle and ReVelle, 1974), attention has more recently been given to the accumulation of mercury in aquatic and terrestrial food chains. Between 1953 and 1960, mercury poisoning (Minamata disease) resulting from consumption of contaminated fish occurred in Japan (Irukayama, 1966) and in the early 1960s Sweden experienced a mercury problem in food. The source for the Japanese illness was industrial mercury waste from a chemical factory, accumulated from sediments and water into fish, which were consumed by humans. In Sweden the source was identified as a mercury fungicide which passed through the food chain from grain to hens to eggs to humans (Löfroth, 1970). The eggs contained 0.029 ppm mercury compared with 0.007 ppm in eggs from other European countries.

For mercury, the chemical form of the metal is all important in recognising the health risk. Organic mercury, especially methyl mercury, is much more soluble and more toxic than inorganic mercury. Methyl mercury has a half-life of 70 days in the human body compared with six days for inorganic mercury. Mercury accumulates in food chains and in addition, biological processes can transform inorganic to methyl mercury (Jensen and Jernelov, 1969). The chemical and biological reactions of mercury in the environment are complex and incompletely understood. Even naturally occurring mercury can be solublized and released into the biosphere as a result of other processes such as acidification of water resulting from acid precipitation.

Detailed studies on mercury in food chains have resulted not only from increased awareness of the potential risks, but also from improved analytical techniques (e.g. Linstedt and Skorfuing, 1972). Studies on fish have shown that mercury is taken up as methyl mercury, either emitted as such, or methylated in sediments by microorganisms. The rate of uptake and the retention time for mercury in fish varies greatly with the species and environmental conditions. It is not yet clear whether the major pathway of mercury to fish is via the water (direct uptake) or from contaminated food (food chain accumulation). Both of these processes can and do occur. However, at the next level of the food chain, when birds prey on fish, or man consumes fish in the diet, food is the major source of mercury intake.

It is tempting to extrapolate from these data and make generalizations for the potential impact of all persistent substances in food chains. Clearly, compounds or elements which exhibit any of the properties listed for DDT should be regarded as potential risks to the health of humans, biota and the environment and thus worthy of close scrutiny. More precise models can however only be constructed on the basis of hard scientific evidence of which there are basically two types:

(1) *Observations and measurements* of concentrations of the pollutant in field situations which give a snapshot at a moment in time but no information on the dynamics of the system — the rates of uptake, loss, retention time etc. at each level of the food chain, nor the toxic effects on populations. Sampling extended in time can partially clarify this, and is also useful for monitoring the disappearance of the pollutant with time.

(2) *Experimental situations* where populations and organisms are isolated and examined for toxicity, rates of uptake and excretion etc., or where simplified ecosystems are set up in microcosms, and known amounts of the substance (often isotopically labelled tracers) are introduced.

Both types of study have great value; ideally the experimental approach should precede the field studies, but this is clearly not possible for pollutants which have already been allowed to escape into the biosphere.

Our awareness of food chains as risk systems should alert us to the need for careful screening of potential pollutant substances. Experimental work in microcosms (experimental artificial ecosystems) are concerned with this aspect of the problem. But we have not always been wise before the event. Where contamination of food chains has already occurred, or is continuing, e.g. from lake sediments which acted as 'sinks' for PCSs, continued monitoring will improve understanding of the complex biological transfers that we are trying to model.

3.6 ESTABLISHING THE RELATIONSHIP BETWEEN THE DOSE AND THE EFFECT

The last link in the food chain, the last piece of any model of a risk system, is the link that connects the impacting variable to its receptor. The importance of the last link justifies the special attention that it frequently receives in the estimation of risks.

What amount of ultraviolet-B radiation, or what increase in the amount is required to produce a specified additional number of cases of skin cancer? In a drought, what level of moisture deficiency and what period of time is required to reduce crop yields significantly or to produce crop failure? How much ingestion of mercury in fish or in bread made from seed-grains is required to produce mercury poisoning? How much damage will result from a flood that covers the floodplain with a metre's depth of flood water, and will the damage be twice as large if the water rises another metre?

These are all examples of the same kind of relationship although the terminology used to describe them differs. For toxic pollutants in the

62

environment or in food chains the term dose-response or dose-effect is used. For droughts the phrase more frequently used is impact-response or impact-effect. If one particular indicator of drought is used the term may be moisture deficit-effect or drought index-effect. For floods the relationships may be expressed in stage-damage curves.

Whatever the terminology used there are important questions to be asked about the curve, its slope and the nature of the relationship. They are important because they have considerable bearing on the setting of standards and the adoption of policies to reduce risk.

3.6.1 Dose-Effect Relationships

Threshold relationships are shown in Figure 3.9. The simple case is in curve (1) where there is no risk until a certain level of exposure takes place. The level of exposure is the point where curve (1) leaves the abcissa.

A more complex and more common case is illustrated by curve (2). Here there is some effect at low doses, increasing relatively slowly, until a take-off point is reached at which the effects increase dramatically. Measured over a

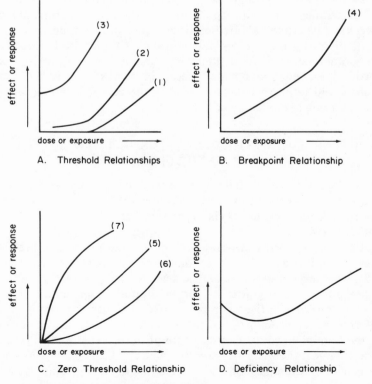

Figure 3.9 Dose-Effect Relationships

whole population this could be a case where a few susceptible people are affected by low exposures but the mass of the population are not affected until a certain threshold or take-off point is reached.

Both these curves may clearly be linked to policy options. In the case of curve (1) it may seem necessary to keep the level of exposure for all persons below the threshold level. In curve (2) it may not be practicable to reduce exposure to zero in order to protect the relatively small number of susceptibles affected. In this case the standards may be set at the take-off point or somewhat below it, and additional steps taken in other ways to safeguard or reduce the exposure of susceptibles.

A more complicated case is shown as curve (3) where the effects of exposure are impossible to separate from similar effects which occur without exposure. In other words there are some 'effects' at zero dose and any dose above zero will increase those effects.

A breakpoint relationship is shown in Figure 3.9(b). In curve (4) a marked break in response occurs at some (usually higher) level of dose.

The curves shown in Figure 3.9(c) are all variations of zero threshold relationships. Curve (5) is the classical linear exposure/effect relationships when no theshold exists. Zero risk occurs only at zero exposure. This is the conservative position that is assumed to exist for exposure to ionizing radiation.

Curve (6) is a variant showing lower sensitivity to risk at lower levels, and curve (7) is the reverse showing increasing sensitivity at lower levels of risk exposure.

In practice it is rarely possible to specify the slope of the dose-effect curve with confidence or to state exactly where the threshold level is if it exists. A number of factors account for the lack of precision in dose-effect curves. The vulnerability of individuals varies for day to day and physiological diversity in human populations is such that effects may vary according to the segment of a population exposed. Measuring techniques have their limits of detection and monitoring of levels can only be carried out in a few selected samples or sites. At very low exposures there may be effects which are not easily detectable and observations at the upper end of the curve are difficult to obtain because massive exposures are relatively rare. A great deal has been learned about radiation exposure of high levels from cohort studies of the populations of Hiroshima and Nagasaki. Normally, ethical considerations prohibit the deliberate exposure of human beings to large amounts of hazard. An alternative approach often used is the conducting of tests on animals but extrapolation of animal data to estimate human experience is always imprecise, and unavoidably so.

The principle use of dose-effect curves is thus to predict the consequences of very high and very low exposures which usually cannot be adquately observed or measured.

In Figure 3.10 a generalized exposure-effect curve shows uncertainty of estimates at high and low levels of exposure and the need to extrapolate

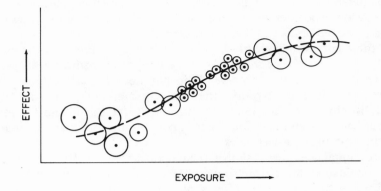

Figure 3.10 Generalized exposure-effect curve, showing uncertainty at high and low exposures: diameter of circles indicates degree of certainty about data points

Source: Lowrance, 1976, p.38. (Reproduced by permission of Wm. Kaufman, Inc.)

upwards and downwards from 'middle range' observations.

As indicated above, the practice in radiation exposures has been to extrapolate to the point of origin — in other words to make a 'no threshold' assumption. In many other areas the conventional approach has been to assume that a threshold level exists and that a small amount of a bad thing is harmless. There is a current trend towards the discovery of harm at lower and lower doses. The threshold concept is on the defensive and many scientists are coming to believe that for many hazardous substances — especially new substances not previously found in the human environment — any exposure is potentially harmful or risky.

There is other evidence however, that some toxic substances are required by the human body in very small quantities for complete health. Deficiencies of some elements, including the toxic heavy metals can in fact be detrimental. In such cases the dose-effect curve would appear as in Figure 3.9(d).

Three different types of tools are used to develop the information needed to construct and interpret dose-response curves: (1) Clinical studies; (2) Epidemiological (population) studies; and (3) animal studies.

Clinical Studies The most reliable subject for determining health effects on man is man himself. For obvious reasons, the use of man as the subject of experiment (clinical studies of harmful impacts) must necessarily be extremely limited and subject to extremely stringent control and safety procedures. There are areas where effects of low exposures can be examined in a carefully controlled clinical study. For example, the US Environmental Protection Agency is conducting a series of controlled human exposure studies to examine the effect of being exposed to low concentration of sulphates.

Epidemiological Studies A second tool which directly provides data on human effects are epidemiology or population studies. These studies focus on following a specific group of individuals over a period of time to see if patterns linking cause and effect can be discerned. The conclusions drawn are necessarily based much more on inference than in the controlled clinical studies where one knows to what the subject has been exposed. There are two types of population studies that are being developed, *retrospective* and *prospective*. *Retrospective* population studies essentially start from an observed effect and attempt to trace back to find what might be the possible cause or causes of the effect. For example, such a study might focus on a geographical area that has a high cancer incidence (high cancer cluster). The study would be designed to detect similar patterns and life style, conditions, or environmental factors, which could have contributed to the observed effect. *Prospective* population studies, on the other hand focus on following populations which have a known exposure to a particular substance, for example, a carcinogen, in order to determine what the possible effects may be to that exposure. Figure 3.11 portrays how such studies fit into a proposed US EPA research Plan to develop data needed to determine Environmental Exposure/Cancer Relationships.

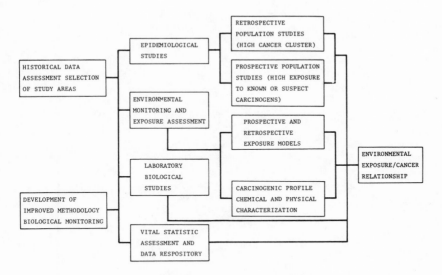

Figure 3.11 Studies being undertaken to establish the dose-effect relationship for environmentally caused human cancers
(US Environmental Protection Agency, 1974)

Animal Studies Most of the information on dose-response comes from animal studies. Tests are run on various animals to determine acute, chronic and genetic effects. Selection of the specific types of animals depends upon the effect to be examined; the target organ that is likely to be affected, and the extrapolation of that response to man; and on the sampling, reproducibility, and timing requirements of the experiment. Extrapolation of animal data to man is the major problem for using this tool and considerable research is needed to narrow the error bar associated with such extrapolations.

CHAPTER 4

Risk Evaluations and National Policy

4.1 POLICY CONSIDERATIONS

In public policy, risk assessments made on the basis of scientific evidence or public alarm, have to be translated into statutes and regulations that can be enforced and, if necessary, stand up in courts of law. National policies have broadly taken two approaches: that of specifying generally applicable codes and regulations based on what is known about the causes and effects of the risk; and a case-by-case approach in which the specific circumstances of each case of the risk is considered individually. The first approach is based directly on the evidence available about causes and effects and levels of harm. The case-by-case approach is, in addition, based on what was *reasonable* in the particular circumstances (sometimes *despite* the harm involved). The criterion of reasonableness has a long tradition in legal systems, so that even when assessments about new risks are enacted in legislation, they are often evaluated according to some very old legal concepts.

Policy decisions about risk are also influenced by the particular circumstances of a hazard or situation being assessed. Among the factors often to be considered by risk assessors are the following questions:

— What alternatives are available, and what are their risks?
— What is the probability of the hazard occurring?
— What are the consequences?
— Who is at risk?
— Are the effects reversible and temporary, or not?
— What is the degree of public concern?

4.1.1 What Alternatives are Available?

The weighing of alternatives is at the heart of the risk assessment process. A contaminated well is not going to be closed down by public health authorities if it is the only water source within miles, whatever national water quality standards have been adopted.

In the control of chemical hazards, the evaluation of alternatives is particularly difficult because often the risks and benefits of the alternatives are as uncertain as the suspect chemical. In more recent decisions, the *risks of substitutes* which will inevitably fill the gap of the prohibited product, are taken more explicitly into account. Thus the proponents and opponents of the asbestos industry argue alike that the risks of substitutes such as glass fibres, which may have similar effects on human health to asbestos fibres, must be part of the overall assessment equation. At the same time, the additional deaths from fires etc., if asbestos is *not* used, needs to be set against the toll of ill health and premature death attributed to the use of asbestos. This view is put forward as a general principle for radiation protection:

> The public must be protected from radiation but not to the extent that the degree of protection provided results in the substitution of a worse hazard for the radiation avoided. (BEIR, 1973)

The importance of having alternatives has two aspects — the availability of substitutes and the freedom to choose. Increasingly where risk assessments result in decisions that intrude into the lives of many people, the issue of individual choice is raised.

4.1.2 Urgency

One factor in the degree of risk is some measure of the imminence of danger. A hazard that is expected soon will require a different evaluation from one whose probabilities are viewed in a longer time frame. The hazard is said to be 'present'. The overhanging slope, the flood, the hurricane, the earthquake, that appear imminent are responded to by protective measures which often include temporary evacuation. When the threatening event has occurred or when its imminence appears over, people's lives return to a more normal pattern; regulations temporarily imposed are lifted, and the assessment of risk is again viewed in a future perspective.

In the USA, pesticides can be prohibited immediately if they are considered to present an imminent hazard. This is defined as

> 'unreasonable adverse *effects* on the environment' meaning 'any unreasonable *risk* to man or the environment, taking into account the economic, social and environmental costs and benefits of the use of any pesticide' (FIFRA, 1972)

This section of the Pesticide Act referring to imminent hazard has been open to considerable dispute since it was first proposed, but it was used by EPA as a basis for immediately suspending, in September 1974, aldrin and dieldrin, two pesticides suspected on the basis of animal experiments to be carcinogens.

Common sense dictates that more imminent dangers require more drastic action than those where there is time for careful risk assessment before taking decisions. The Hippocratic rule of medicine that extreme remedies are appropriate for extreme diseases echoes what a reasonable person would do. Public health and other government authorities are more willing to allow doctors to use less thoroughly tested drugs where threats to life are imminent. However, for many hazards, especially chemical ones, the issue is not just what action is required when the danger is imminent, but how to define what the indicators of 'imminence' are — in other words what are the early warnings we should take notice of?

4.1.3 Accidents

Events take on a particular significance when they are rare and unexpected. The example of the explosion in a chemical plant at Flixborough in England in June 1974 illustrates how one accident can completely change the risk assessment by making an 'incredible event' only too credible. Flixborough has been described as the single most important event in forming British public opinion about industrial hazards since the second world war (McGinty, 1976). It is the archetype of industrial disasters to be avoided and has been the subject of an important investigation by the UK Health and Safety Commission's Advisory Committee on Major Hazards (1974).

The report reveals that inadequate consideration was given in the design and management of the chemical plant to the *possibility* of the accident happening. In effect, there was no risk assessment made for the event and thus no judgement made on its acceptability, either in consequence or probability terms. Thus a very different situation existed in the UK for major chemical plants compared to the detailed risk assessment and stringent regulations in safety design and management that are imposed on nuclear power stations.

4.1.4 Other Factors

In industrialized countries, there now exist many case histories of risk assessment that are well documented. One important collection of these for the USA is concerned with forty-five cases of public concern and government

Table 4.1 Factors Leading to Prompt Public Policy Action

Risks are judged to be serious and action is more likely to be taken when:

— a serious event has just occurred
— there is a high probability of a dangerous occurrence
— danger is imminent

— the danger is new and unfamiliar
— the danger has other fear associates (e.g. nuclear bomb)
— the danger is carcinogenic, mutagenic or teratogenic

— the damage is acute and short term
— the damage is irreversible
— many people are potentially affected
— cases of harm occur together (at one time or place)

— children are affected

— cause-effect relationships are scientifically understood
— there is direct impact on people (i.e. not through long cause-effect chains)
— national security is involved
— it is publically known
— it is given much attention in the mass media and by politicians
— consequences are highly damaging to economic and trade interests

— alternatives are available

---

response for hazards which range from DDT, mercury discharges from industry, massive oil spills, athletic injuries, and polio vaccine, to x-ray machines and oral contraceptives (Lawless, 1977). Many European case histories are documented in scientific journals like *Ambio* (Sweden) and *New Scientist* (UK). From these case histories some general findings emerge about the factors likely to produce an assessment that concludes the risks are serious and warrant immediate action, compared to others that are more likely to lead to administrative delay while further evidence is forthcoming. Some of these factors are summarized in Table 4.1. It should be emphasized that these do not necessarily represent guidelines for future decision-making, but rather the priorities, trade-offs and social judgements associated with present risk management decisions in certain industrialized societies. They describe present processes rather than a model of how risk assessment should operate.

4.2 LEGISLATIVE CONSIDERATIONS

In Chapter 2, risk assessment was shown to be a comparative process, once the hazard itself had been defined (Figure 2.3). In legislation, these comparisons are made into explicit criteria for deciding whether a hazard is acceptable or not. Figure 4.1 illustrates how public policy judgements about risk are based on the different comparisons that can be made. These judgements are often enshrined in particular phrases, which appear in the legislation itself. In many countries, the exact meaning of these phrases is spelled out, not in the legislation, but in the courts. Increasingly, the judiciary of many countries, most notably in the USA, is taking upon itself the role of clarifying and interpreting the intentions of legislators in drafting environmental policy.

Countries in which detailed codes, standards and regulations are drawn up to make a risk assessment part of public policy, place heavy reliance on the scientific evidence available *within the risk system itself*. The rules are usually intended to be across-the-board and apply to all cases. Thus if scientific evidence is accepted which shows evidence of adverse effects of a pollutant at certain concentrations, then these concentrations (with a safety factor) will be the basis for specific standards. Any person or agency causing these standards to be exceeded can be automatically dealt with under the terms of the legislation.

Where a comparison of the risks with the benefits is formally a part of the legislative assessment, then rigid codes are replaced by judgements framed within the context of individual cases. Two widely used approaches in national policies apply one of two questions to each case:

— What is a practical solution? (best practicable means)
— Are effective benefits being produced? (efficacy)

In both these approaches, public policy is adopting a risk-benefit approach. In the case of 'best practical means' very high benefits can sometimes outweigh the risks. For example, a factory with high pollution levels may be tolerated

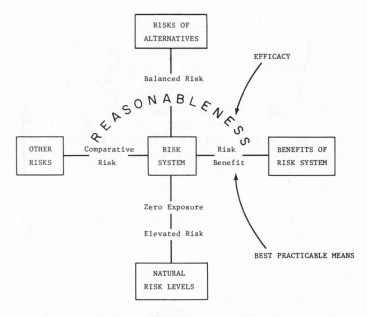

Figure 4.1 The risk comparisions contained in legislative principles

because it brings needed employment to an area and it would have to close down if rigid effluent standards are enforced. On the other hand, for factories where emission levels can be improved without a significant loss of benefits, then a 'best practical means' approach can lower emission levels *below* a general standard. A risk-benefit approach can thus be viewed as risk assessment on a sliding-scale in contrast to uniform standards and regulations. Either approach can produce more or less strict standards and controls. Indeed, public policy decisions often result from successive uses of different criteria as one or another prove unsuccessful.

The role of hazard classification becomes therefore very significant in defining the 'rules of the game'. Specific instruments of public policy deal with categories of risk rather than all risks. Thus nuclear installations, major industries, earthquakes (building codes), water pollution (effluent controls), pesticides, drugs, etc. all come under separate public policies often administered by different agencies. Since criteria for risk assessment are embodied in public legislation they also differ for different categories of risk. A reclassification of a risk (e.g. from being a food additive to a drug; from being an ordinary industrial undertaking to a 'major works'; from being outside a flood zone or earthquake zone or clean air zone to being inside) may well mean that a different public policy criterion will be applicable in assessing the risk. Recent legislation (e.g. the US Federal Resource Conservation and Recovery Act of 1976) can be seen as a process of rationalizing risk assessment criteria across the risks entering the man-environment system through its air, water and land components.

The process of making risk assessments consistent with one another is a major theme in the evolution of public policy. In some countries it is being approached by increasing the codification of risk assessment — by the specification of more rules, regulations, standards and procedures — and also by increasingly quantitative formats for assessing risk.

What has emerged so far from this survey of public policy for risks is that an increasingly quantified and codified assessment process is not necessarily based on new legislative principles. Rather, the concepts of 'best practicable means' and 'reasonableness' remain so far the fundamental tools for the decision maker in many countries to evaluate the evidence and arrive at a (defensible) decision. The US Toxic Substances Control Act 1976 and the UK Nuclear Installations Act 1965 are only two examples where these concepts underlie highly technical codes.

4.2.1 Scientific Criteria

Criteria are observed effects/responses of one or several contaminants on a defined reception/population under specific conditions. For air pollution, for example, such criteria are used, along with socio-economic considerations, as the basis for establishing *air quality standards*, which are maximum allowable concentrations for given times and contaminants (Munn, Phillips and Sanderson, 1977).

However, scientific criteria are not the undisputed canons that they are sometimes believed to be. Different scientists, and thus different government agencies, disagree about what constitutes a *relevant, valid and acceptable* basis for assessing risks. These differences arise because the scientific data, especially for new hazards, is less than ideal and contains gaps in knowledge about the risk systems, and about cause-effect relationships. Two issues often lie at the heart of the differences: that of what constitutes a *significant* effect, and what is an *adequate amount of data* on which to make a scientific assessment of risk.

(A) *Adequate data* Scientific data on a suspected pollutant or risk rarely all point in the same direction. Experiments on one animal species on the toxicity of a chemical will not necessarily have the same results as similar experiments on another animal species. Epidemiological evidence frequently appears to contradict, or at least fails to confirm for human populations, the toxicological data obtained from animal experiments. Sometimes proof of carcinogenesis is *conventionally* accepted when found in tests of two or more species. At other times, decisions are taken on the evidence from experiments on one species. Thus what are 'adequate' data will differ from country to country, from time to time, and from risk to risk. Adequacy is highly bound by its context and may stray within and without the norms of scientific convention.

(B) *Significant effects* In the USSR, the present criteria for atmospheric pollutants derive from the principles laid down by V.A. Rjazanov in 1949. The criteria are:

> (1) The only permissible atmospheric concentration of a given substance is that which causes no direct or indirect, harmful or unpleasant effect in man and which does not impair either his ability to work or his mental or physical well being.
> (2) Habituation to harmful substances must be considered as a harmful sign and is proof that the concentration in question is above the permissible level.
> (3) Other impermissible concentrations are those at which atmospheric pollutants have a harmful effect on vegetation, local climate, atmospheric visibility, or human living conditions.' (Izmerov, 1973)

By these criteria therefore 'significant effects' include direct *and* indirect, harmful *and* unpleasant, physical *and* mental effects. They also include *adaptive* responses to changed environmental conditions and effects on the environment itself. This comprehensive definition of 'significant effects' is consistent with experimental design in the USSR in which changes in behaviour and biochemical processes are tested. In particular Soviet scientists have developed studies to detect compensatory or adaptive responses for detecting the effects of long term, low concentrations of pollutants. In addition to experiments they use questionnaires on living conditions and special registers to monitor incidents such as minor eye injuries from pollutants.

In this regard, the Soviet definition of what constitutes a significant effect seems to be more stringent than that accepted in many other countries. It falls near the lowest level of effects in the WHO (1964) categories for atmospheric pollutants.[1] In other countries such as UK, USA and Canada, adaptive responses of an organism to a pollutant are not defined as 'significant effects'. The issue of low concentrations of lead in human blood is one example of different national standards being equally based on scientific criteria of dose-response relationships but interpreting the significant cut-off point for harmful effects at different levels of response. In this case, the effects concern the activity of one enzyme in the haemoglobin chain (Hernberg, 1972).

The number of situations in which the above uncertainty about significant effects exists, is increasing as better measurement techniques can reveal more subtle and sub-clinical biological effects. Our ability to interpret their significance lags behind our measuring facilities and gives rise to a wide range of national interpretations of what scientific criteria are and what they mean.

This problem is widely acknowledged by scientists involved in risk assessment. For example, in the USA, the Surgeon General's Advisory Committee on Smoking and Health, 1964, reported on scientific criteria as

> Statistical methods cannot establish proof of a causal relationship in an association. The causal significance of an association is a matter of judgement which goes beyond any statement of statistical probability. To judge or evaluate

the causal significance of the association between the attribute or agent and the disease, or effect upon health, a number of criteria must be utilized, no one of which is an all-sufficient basis for judgement. These criteria include:

a) The consistency of the association
b) The strength of the association
c) The specificity of the association
d) The temporal relationships of the association
e) The coherence of the association

Public policy legislation can help to mitigate this uncertainty by requiring risk assessments to be based on several sources of evidence. For example, the air quality criteria recommended by a WHO Expert Committee in 1972 include evidence from experimental animal studies and effects on vegetation, as well as effects on man (acute and chronic effects, annoyance reactions, effects on athletic, cardiovascular and psychomotor performance).

National drinking water standards are commonly based on several criteria, including physical, chemical, radiological and microbiological characteristics. Some of the measures involved are based on scientific experimental and epidemiological evidence. Others, such as those relating to colour, taste, turbidity, and odour of water and nuisance organisms are subjective criteria defined in terms of people's acceptance or rejection of water rather than necessarily being health risks themselves. These include physical characteristics (indicators of organisms) and coliform bacteria (indicator of enteric pollution present).

Thus the scientific criteria developed for regulating drinking water quality can be seen as a mutually reinforcing and comprehensive system of risk assessment which includes objective, subjective and indicative bases for judgement. However, for most hazards today, the risk assessment system is not as soundly based on scientific criteria as it is for drinking water quality.

Codes which can be based on scientific criteria tend to reduce the initial emphasis in the risk assessment on a case-by-case approach. Regulations and coded procedures seek to generalize risk assessment across individual situations. Ultimately if the assessment becomes controversial, and particularly if it is contested in courts, the general applicability of the codes may break down. Courts tend to interpret the behaviour of any individual or organisation involved in risks according to the criteria of 'reasonableness' for a particular context. Conformity with codes may or may not limit the liability of an individual, but it will not absolve the whole risk assessment system. Thus a manufacturer may be indemnified against blame because he can show that regulations were followed according to the letter of the law. The liability may then shift to the regulating agency who drew up the procedures. In any case, somewhere in the system, responsibility (and liability) has to rest, whether decisions are made too early or too late; made, revised, or not made at all.

4.2.2 The No Risk Principle: Zero Exposure

The principle of zero exposure to a particular risk may seem at first to cut a bold line in public policy. Its guiding principle is that no level of risk is acceptable from the hazard in question. As a criterion stated without qualification, it does not require other factors, such as the availability of alternatives, to be taken into account. Taken literally, a criterion of zero exposure would exclude consideration of other criteria commonly used in public policy. It thus stands apart from 'best practicable means', 'reasonableness', and 'efficacy' which are implicitly part of a risk-benefit assessment. Zero exposure puts the emphasis on the 'risk' side of the equation.

Zero exposure is an issue in the present debate about nuclear power stations and particularly about plutonium economy and nuclear waste reprocessing in North America and Europe.

In present public policy, zero exposure is most used in assessing the risks from suspected carcinogens. Experience in the USA particularly through the well-known Delaney Clause and the less well known 'Cancer Principles' is the best documented attempt to use a zero exposure criterion. In the UK, a zero exposure principle is used in the Pesticides Safety Precaution Scheme under which British companies voluntarily agree not to sell pesticides which have not been cleared as non-carcinogens.[2] The American experience also illustrates the difficulties and many qualifications necessary in applying such a criterion in risk assessment. Far from being an effective sword for public policy, 'zero exposure' has proved to be a two-edged one in bringing criticism down on the heads of agencies who have tried to implement it.

One difficulty with using a zero-exposure or zero tolerance approach in legislation is that the definition of 'zero' depends on what measuring techniques and accuracies are available. In earlier legislation, zero tolerances have been set as a matter of convenience when tests did not reveal any residues. Legally, it meant that no residue was *permitted,* but technically it meant that no residue could be *detected* (Blodgett, 1974). What happened when the techniques for measuring residues improved by several orders of magnitude, as they did in the 1960s, was that previously 'uncontaminated' material was suddenly declared contaminated. Today's analytical techniques are almost one *million* times more sensitive than they were in 1958. For example, in the USA, 50 parts per million was regarded as the practical equivalent of 'zero' in 1958. Today, electron microscopes, mass spectroscopy, neutron activation analysis and gas chromatography can collectively detect amounts of contaminants as little as *one part per trillion* (10^{12}).

In the US, the Department of Agriculture, which has registered pesticides on the basis of zero tolerances, had to establish a fund to indemnify farmers whose crops were suddenly seized because they contained residues that could be detected with new analytical techniques. Since the farmers had sprayed the pesticide as recommended by the Department of Agriculture, the authorities were also held responsible. This type of situation arose because, although zero tolerance is in fact dependent on the best available technology for measuring residues (that is, is a relative concept) when framed in legislation it has been defined in absolute terms.

76

The Delaney Clause and the Cancer Principles in the USA The American experience with zero tolerances in legislation is one of the most widely discussed internationally and has been compared favourably and unfavourably with legislation in other countries. The best known legislation is the Delaney Clause[3] which refers to food and colour additives. It states that an additive:

> shall be deemed unsafe, and shall not be listed for any use which will or may result in ingestion of all or part of such additive, if the additive is found by the Secretary to induce cancer when ingested by man or animal or if it is found by the Secretary, after tests which were appropriate for the evaluation of the safety of additives for use in food, to induce cancer in man or animal ...

The Delaney Clause was invoked in 1969 for the banning of cyclamates and in 1977 for the banning of saccharin. In both cases the use of the Clause has provoked a storm of protest over the 'zero exposure' criterion. Subsequent FDA consideration for reclassifying the chemicals as 'drugs' rather than 'food additives' has also produced debate about the inconsistency of the criteria used for risk assessment.

The arguments *against* the Delaney Clause are that it is too rigid and does not allow case by case assessment. It is an 'all or nothing' criterion in decision-making. A panel of the President's Advisory Committee objected (1973) to the Clause because

> The rigid stipulation of the Delaney Clause, springing from presently inadequate biological knowledge, places the administrator in a very difficult interpretative position. He is not allowed, for example, to weigh any known benefits to human health, no matter how large, against the possible risks of cancer production, no matter how small.

A second argument against the Delaney Clause is that the ability of detection today to one part per trillion (10^{12}) far outstrips our ability to *evaluate the significance* of such minute amounts, even of carcinogens.

Proponents of zero tolerance argue however that there is *no* safe level for carcinogens given present scientific knowledge about dose-response relationships in cancer-producing agents. However, it is also known that common salt, and indeed, food itself, can cause cancer when taken in large enough amounts.

The application of a zero exposure criterion for carcinogens is being extended in the USA to chemicals entering the environment and food through pesticides.

From 1973 onwards the US Environmental Protection Agency has developed a set of so called 'cancer principles' for assessing the risks of potential carcinogens which come under EPA's jurisdiction (Table 4.2). It is part of an ongoing effort to rationalize risk assessment for suspected carcinogens across the jurisdictions of different agencies — in other words, to overcome the inconsistencies created by a hazard taxonomy that divides chemicals into drugs, cosmetics, food additives and pesticides etc.

By June 1975 nine principles had been expanded to seventeen. The purpose of setting down the principles in these hearings was to put them into the record

Table 4.2 Nine Cancer Principles of US Environmental Protection Agency (1974) (Reproduced by permission of the US Environmental Protection Agency)

---

(1) A carcinogen is any agent which increases tumor induction in man or animals.

(2) Well-established criteria exist for distinguishing between benign and malignant tumors; however, even the induction of benign tumors is sufficient to characterize a chemical as a carcinogen.

(3) The majority of human cancers are caused by avoidable exposure to carcinogens.

(4) While chemicals can be carcinogenic agents, only a small percentage actually are.

(5) Carcinogenesis is characterized by its irreversibility and long latency period following the initial exposure to the carcinogenic agent.

(6) There is a great variation in individual susceptibility to carcinogens.

(7) The concept of a 'threshold' exposure level for a carcinogenic agent has no practical significance because there is no valid method for establishing such a level.

(8) A carcinogenic agent may be identified through analysis of tumor induction results with laboratory animals exposed to the agent, or on a post hoc basis by properly conducted epidemiological studies.

(9) Any substance which produces tumors in animals must be considered a carcinogenic hazard to man if the results were achieved according to the established parameters of a valid carcinogenesis test.

Note: These principles were followed by 29 pages of citations from reports and testimony to support each principle and to place them in context.

and to establish their validity as *criteria* for public policy through their incorporation into Administrative Law decisions. The cancer principles were distilled from the testimony at the various pesticide hearings of many scientific and medical experts. Each principle was supported by detailed documentation.

The use of these 'cancer principles' as legal facts and scientific criteria in EPA decisions to suspend or cancel pesticides (notably the aldrin and dieldrin suspension in 1974, DDT in 1972, and the heptachlor and chlordane proceedings in 1975) has caused much controversy among US scientists and decision-makers.

The first principle (a carcinogen is any agent which increases tumour induction in man or animals) caused controversy in equating the production of any tumour — benign or otherwise — with carcinogenesis. It was regarded as a precedent-setting criterion, although it was similar to the Delaney Clause. EPA also felt the burden of proving safety should be on industry and not on the public, and recommended the use of experimental animals such as rodents which were *more* sensitive to toxicity than man, following a philosophy of better to err in the direction of preventing a 'safe' substance getting into the environment than to err in allowing the entry of a harmful one.

Although the use of the cancer principles extends present FDA policy on food additives over a wider range of chemicals in the environment, one major source of opposition to the principles is from within FDA and from the food and drug industry which oppose the Delaney Clause. They would seek to rationalize the US national risk assessment policy to chemicals on another basis than that of 'zero exposure'.

4.2.3 Best Practicable Means

The term 'best practicable means' is widely used in environmental legislation in English-speaking countries, both industrialized and developing. It has been described (Ashby, 1975) as 'an elastic band ever tightening as chemical science advanced' and has operated as a criterion for regulatory control and court judgements in many countries. It is commonly though of as a peculiarly British institution, where is is used to determine effluent standards from major industrial works and to control air and water quality generally. The same, or similar criteria for regulation and liability operate in other countries, including Australia, Canada, Federal Republic of Germany, Jamaica and the USA.

Related terms include:

(1) *best practicable technology* or *best practicable control technology currently available* (BPT), e.g. 1972 Amendments to the US Federal Water Pollution Control Act.

(2) *best available technology economically achievable* (BAT), e.g. 1972 Amendments to the US Federal Water Pollution Control Act (1983 standard).

(3) *any practicable means* or *all practicable means,* e.g. UK Clean Air Act 1956 (provisions for smoke emissions from railway locomotives), and Clean Air Act 1957 State of Victoria, Australia.

Although the term (and its variants) are widely used, they are usually not clearly defined within the legislation where they appear. This flexibility in interpretation is often cited by proponents of the term as its major advantage, and by opponents as its major weakness.

The Clean Air Law 1961 of Jamaica is one of the few that include a definition of 'best practicable means'. The word 'practicable' is defined by the Act to mean

> 'reasonably practicable, having regard, amongst other things, to local conditions and circumstances, to the financial implications and to the current state of technical knowledge.'

This definition is similar in wording to that of the UK Clean Air Act of 1956. In both Acts the importance of *case by case* assessments, the *current* state of knowledge and the weighing of *economic and technical* factors are made explicit. However, other considerations, such as environmental impact and aesthetic quality are not excluded from the risk assessment just because they are not specified. Decisions by regulating bodies and individual inspectors, and judgements made in court have interpreted the term 'best practicable means' within a wide range of possible meanings.

The distinction to be made between *best practicable technology* and *best available technology* is also not always clear in national legislation. For example, in the US Federal Water Pollution Control Act 1972 Amendments the risk-benefit equation for best practical technology is defined as:

> Factors relating to the assessment of *best practicable* control technology currently available ... shall include consideration of the total cost of application of technology in relation to the effluent reduction benefits to be

achieved from such application, and shall also take into account the age of equipment and facilities involved, the process employed, the engineering aspects of the application of various types of control techniques, process changes, non-water quality environmental impact (including energy requirements), and such other factors as the Administrator deems appropriate. (Sec.304(b)(1)B)

The instruction about how the 'best available technology' is to be assessed is almost identical including the loophole phrase 'and such other factors as the Administrator deems appropriate'. This last phrase would seem able to take into consideration the possibility of best available technologies being found outside the USA and necessitating foreign imports — an issue which has been widely discussed. The requirement to consider the 'best available technology' in 1983 would appear mainly to call for a re-evaluation of control technologies at that time.

The term best available technology (and its variants) leaves several questions open when in legislation. These include:

(1) *What factors are to be included in the assessment?* — technology, economics, material and non-material disbenefits (such as scenic loss).
(2) *From whose point of view is practicality to be defined?* — the polluter or the public?
(3) *Who defines what is practicable?* — the polluter, the regulator or the public?
(4) *Does the 'best practicable means' include the extreme case of prohibition* of the cause in order to reduce pollution to zero?

One reason why a 'best practical means' approach has been widely adopted in national legislation is that it can be interpreted to suit national risk assessment philosophies. The degree to which a risk is evaluated in terms only of its impact on public health; or whether it will be broadened to include impacts on environmental quality, is left up to the regulating agency.

The use of 'best practicable means' in risk assessment places a burden on the regulator, and particularly (as in the UK) on the individual inspectors of a regulatory agency. In effect they are required to make political decisions in which they must discriminate between large and small works, or old works with outdated equipment and new ones with modern technology. They take into account the need for jobs locally, the importance of the product, and financial and managerial capability of a firm, and its 'good will'. Such inspectors have been described as

> ...human 'computers' ... for social welfare, programmed by inputs of hard data and social values, ... authorized to include their own judgement as one of the inputs ... entrusted with political decisions ... (Ashby, 1975)

The role of the courts in many countries has been to interpret what 'best practicable means' are. In the UK, a polluter can use the fact that he used the best practical means to control pollution as a defence in any court proceedings taken under the Clean Air Act 1956. Sometimes the courts condemn the use of such terms as 'so far as is practicable' and 'where feasible' in legislation on the grounds that they are too vague but generally they are regarded by judges as providing 'reasonably certain' criteria. Courts in English-speaking countries have also differed in from whose point of view practicality is to be defined. Often, however, it is defined from the standpoint of the polluter or source of risk, rather than from that of the public's welfare. In these cases the risk assessment may err on the side of risk perpetrator.

It has been known, however, for English courts to consider 'practical feasibility' under the rubric of 'inevitability of nuisance' (e.g. UK House of Lords 1930). The significance of this is that an action of nuisance is much easier to bring against a polluter. It is a common law doctrine providing a broad avenue of legal action which requires comparatively light proofs, and which is being used increasingly in environmental litigation brought under Common Law.[5] Many statues relating to public health and environmental pollution declare specific acts to constitute 'nuisances' and thus can be tried on a case by case basis under the common law of nuisance.

'Best practicable means' are sometimes regarded as a regulatory criteria distinct from quantitative probability methods of risk assessment and from the setting of quantitative or specific standards. The UK Nuclear Installations Inspectorate have long argued that their 'best practicable means' rationale have enabled them to press for progressively higher standards of design and operation in nuclear reactors whereas a probability risk assessment might lean towards a less than optimal standard (Critchley, 1976). In the USA individual permits for effluence discharge are arrived at through the 'best practical control technology rationale' (Federal Water Pollution Control Act 1972 Amendments). A standard was also issued for vinyl chloride in the workplace on the basis of what is 'technologically feasible' by the USA Occupational Safety and Health Administration in 1974. This is a particularly interesting use of the 'best practical means' philosophy, as vinyl chloride is a carcinogen.

A quantitative approach to 'best practicable means' is used by industry in the UK for accident risks to workers and to the public from chemical works. This involves estimating the Fatal Accident Frequency Rate for a plant (defined as the number of deaths in every 10^8 hours exposed to risk) compared to the rate for the chemical industry as a whole. Plants which are found to have higher accident risks are then improved 'by the best practicable means' — usually on the basis of an economic analysis of the cost of an accident against the cost of preventing or reducing it (Gibson, 1976).

The use of 'best practicable means' in risk assessment is widespread today and has long historical precedence. It is a flexible criterion which at its best can be used as a lever for the highest standards today, and be forward looking to future improvements. It requires technically competent and ethical personnel within the regulating agencies to define what are the best practical means in each case. It also places upon those regulators the task of almost the whole risk assessment process in miniature, for it is they who balance tangibles and intangibles in case by risk equations. It can be argued that an assessment process employing 'best practicable means' pre-empts some of the judgements that should be made by the public or by their elected representatives. This last argument is particularly an issue where a 'best practical means' strategy is associated with government and industrial secrecy. What the means are, and how effective in controlling the risks, can be withheld from the public at risk behind a cloak of 'best practicable means'.[6] However, the two issues should be clearly separated since the use of 'best practicable means' does not necessarily require confidentiality about the standards set or the means by which they are achieved.

4.2.4 Efficacy

Efficacy is defined as a
(1) Capacity to produce effects;
(2) Power to effect the object intended.

In the context of risk assessment, efficacy assumes a more *relative* meaning, in that how effective or useful a product or process is must be weighed (among other things) against how safe it is judged to be. Efficacy and safety are thus two interdependent concepts in risk assessment although they may be measured or estimated in independent processes and by different bodies. The balancing process has been made particularly explicit in drug safety:

> No physician, no one who has ever been responsible for the welfare of individual patients, will accept the idea that safety can be judged in the absence of a decision about efficacy. No drug is 'safe' if it fails to cure a serious disease for which a cure is available. No drug is too dangerous to use if it will cure a fatal disease for which no other cure is available. To attempt to separate the two concepts is completely irrational. (US Senate, 1960).

Another set of risks for which efficacy is widely used as a criterion is for pesticides. It is explicitly stated in the pesticide legislation in Australia, Finland, France, Japan, Korea, Netherlands, Norway, Sweden, Switzerland, and Taiwan, among others. Field trials are usually required in experimental plots over at least one year, together with controls, in order to test the effectiveness of the pesticides.

Efficacy is also widely used in national legislation as a criterion for controlling consumer product labelling and advertising to counteract false or misleading claims. In some cases, the product was effective but not in the specific conditions for which it was sold, and so it could be banned. For example, in 1964 the US Federal Food and Drug Administration stopped the sale of 50 cold compounds and lozenges containing penicillin and other anti-biotics because they were ineffective against the respiratory virus infections for which they were prescribed, although they are effective in other circumstances. Their rationale was published as

> There is a *lack* of substantial evidence that the drugs are efficacious for the purpose *claimed in the labelling* (Federal Register, 1964).

Efficacy is to some extent always a consideration in risk assessment but it is commonly not made explicit in public policy, even where it might be expected. For example, British legislation requires drug manufacturers to prove that new drugs are safe, but not that they are also effective. On the other hand, efficacy is also used in legislation to regulate claims about products that do not necessarily relate to concerns about safety.

4.3 LEGAL CONSIDERATIONS

4.3.1 Reasonableness

The concept of reasonablesness is very important to risk assessment in the public policy of many countries, particularly those influenced by the English Common Law system. It is frequently mentioned in environmental legislation, and is even more important in the way courts interpret legislation. It lies at the foundation of much risk assessment, even when it is not spelled out. Thus *scientific criteria, zero tolerance* and *best practical means* are all evaluated in the light of what is *reasonable*. A judgement about reasonableness also lies behind many of the technical codes in regulating hazards.

In English, reasonableness means 'what is appropriate or suitable to the circumstances'. Court interpretations of the word as it appears in environmental legislation put it in the context of the ordinary, average man. Reasonableness is that which is

> ... in the scale of prophecy or foresight of the reasonable man ... (Harper and James, 1956)

While this may seem a circular argument, the notions of 'ordinary', 'average', 'natural' as yardsticks for measuring risk and harm are well established in national environmental policies. They enable, for example, an industrial management to argue that their actions were *customary* technical or professional practice; that, is reasonable, acceptable practice. Similarly, under English Common Law a defendant can argue that environmental damages are the result of 'reasonable use' of land where the uses are ordinary for the time and the place. The escape of noxious fumes, on the other hand, has generally been regarded as an 'unreasonable' use of property. This emphasis on the average man in the measure of reasonableness makes it difficult to argue the case of people who are particularly sensitive to environmental pollution, as has been found in the cases of noise and smell.

Both judicial and legislative uses of 'reasonableness' focus on a *case by case assessment* in which the parameters of the particular situation are crucial to the decision. This emphasis on the *context* in which each risk assessment is made imparts an individualized, flexible and, some would argue, a piecemeal approach to the risk management process.

Legislative definitions of 'reasonableness' are hard to find, although in some countries almost every public health statute includes the term. It is usually left to the judicial process to decide what is reasonable for each individual case. The US National Commission on Product Safety has provided guidelines on reasonableness in relation to products:

Risks of bodily harm to users are not unreasonable when consumers understand that risks exist, can appraise their probability and severity, know how to cope with them, and voluntarily accept them to get benefits that could not be obtained in less risky ways. When there is a risk of this character, consumers have reasonable opportunity to protect themselves; and public authorities should hesitate to substitute their value judgments about the desirability of the risk for those of the consumers who choose to incur it.

But preventable risk is not reasonable

(a) when consumers do not know that it exists; or

(b) when, though aware of it, consumers are unable to estimate its frequency and severity; or

(c) when consumers do not know how to cope with it, and hence are likely to incur harm unnecessarily; or

(d) when risk is unnecessary in . . . that it could be reduced or eliminated at a cost in money or in the performance of the product that consumers would willingly incur if they knew the facts and were given the choice.

Much of the history of risk assessment has been a response to events, or a case by case analysis of what has happened. Reasonableness has long been a *post facto* criterion. In recent legislation (notably the US Toxic Substances Control Act 1976) (TOSCA) it is being expressed a new way — future probability. Thus TOSCA uses the criterion as 'reasonable likelihood of future harm'. This is an important new direction which may lead to probability expressions of reasonableness that do not depend upon the circumstances of individual assessments.

Thus reasonableness may itself become codified, although exactly in what way and how it would relate to legal interpretation of legislation, is unclear. Indeed, the other criteria described here can be regarded as earlier attempts to codify reasonableness. If they are, they have not been very successful in moving risk assessment away from case by case analysis of what is reasonable risk and reasonable behaviour and policy. Where decisions are contested, the assessment often stands or falls on what constitutes reasonableness in *that particular case* rather than as it is embodied into existing regulations. In the USA, for example, more and more assessments are being contested in court despite increasing codification of reasonableness through other criteria in regulations.

4.3.2 Nuisance and Negligence

The criterion of reasonableness also underlies other legal concepts such as *nuisance* (continuous or repeated unreasonable acts) and *negligence* (absence of reasonable care) in English Common Law, and *les obligations de voisinage* (reasonable use of property in relation to one's neighbours) in the French Civil Code. In English Common Law, cases brought for negligence and nuisance constitute the main judicial process for risk assessment and acceptance of risk liability. As a separate basis for legal liability, negligence rose with the Industrial Revolution when a rapid increase occurred in the number of industrial and transport accidents. Standards of negligence largely determined the levels of safety required in industry at that time (Prossner, 1964).

4.3.3 Assumption of Risk

Another legal inheritance of industrialization is the common law doctrine of 'assumption of risk' that can be used by an employer or owner against a charge of negligence. Its doctrine, in the words of a Scottish court in 1863, is that 'if a servant in the face of a manifest danger chooses to go on with his work, he does so at his own risk, and not at the risk of his master' (Wollan, 1968).

It allows employers to provide lower levels of safety for their workers, and in many countries, including the USA and the UK, the doctrine has been abolished as a defence in specific circumstances by statute or by regulations such as govern worker's compensation boards. In the USA it was seen as a common law response to the needs of industrial progress.

> Assumption of risk is a judicially created rule which was developed in response to the general impulse of common law courts ... to insulate the employer as much as possible from bearing the 'human overhead' which is an inevitable part of the cost — to someone — of doing industrialized business. The general purpose behind this development in the common law seems to have been to give maximum freedom to expanding industry (9). Justice Black (1942) in Wollan, 1968.

The doctrine is not, however, entirely removed from the statute books of many countries, so that an employer can still argue that an employee assumes those risks which are 'ordinarily and normally incident to work' (reasonable risks?) as well as any extraordinary risks which he *'knows and appreciates* and faces without complaint'.

4.3.4 Professional Liability

In the field of professional liability a similar process of assessing responsibility for risks incurred has evolved which again hinges on judgements of what is reasonable. Historically, professional services (of engineer, architects, builders etc.) were limited in liability by two doctrines in English common law; *'privity'* which restricted liability only to those with whom a contract had been directly made (and thus excluded rights of compensation to the ultimate purchaser or secondary middlemen); and the *'completed and accepted'* rule which said that once the owner or purchaser accepted the work or product then he took all responsibility for it, including risks to third parties arising from defects.

These two rules have been criticized and changed (particularly in the USA) so that by the 1960s neither manufacturers nor professionals were protected from responsibility for risks arising from their goods and services. The burden for these risks swung too far in the other direction in the USA, and professionals found themselves liable on negligence charges for indefinite times into the future in connection with projects they had virtually forgotten and a changed climate of what constitutes 'good practice' and 'reasonable prudence and competence'. They could not realistically assess the *total future risks* they were making in decisions about a project where their liability was unlimited in time.

The unreasonable burden this placed upon them was increasingly recognized in the late 1960s in the USA when many states enacted legislation to limit the number of years professionals could be exposed to liability for negligence on a particular project. In California, for example, a four year time limit was made from the date of 'accepted and completed' by the owner. In other states, the limit ranges from three to ten years. The situation for professional liability in the case of medicine is different; in some states, patients can bring proceedings for a limited time period after he (the patient) discovers the negligence or malpractice, which may be 20 years after it actually occurred. The judicial trend is to date the time-limit from the point of discovery of damage rather than the date of 'acceptance and completion' (Fife, 1973).

When cases concerning risk and liability reach the courts there are several general factors that presently influence judicial decisions about what constitutes a reasonable risk. These include the probability of harm, the magnitude of the consequences, the degree of knowledge, the availability of alternatives and the ratio between the risks and the benefits involved.

4.3.5 Probability of Harm

Legislative and legal approaches to 'reasonableness' consider how likely an accident or injury is in the eyes of an average, prudent person. Thus some events are regarded as being unavoidable in that they have probabilities below that at which an average person would consider taking precautions. They would include a runaway horse and an unprecedented natural hazard such as a flood or lightning strike. For these hazards, the law of negligence would not apply and no one would be legally responsible.

4.3.6 Magnitude of Consequences

The criterion of reasonableness itself constitutes a judgement in which probability and consequence are combined. Thus, when the consequences can be expected to be great (such as causing death), a reasonable person would be (or should be) more prudent even when the probability is low. For example, shooting a gun blindly in a forest may have a low probability of hitting someone, but the shooter would be considered negligent or 'unreasonable' in doing it because the consequences could be serious. In common law,

> As the gravity of the possible harm increases, the apparent likelihood of its occurrence need be correspondingly less. (Prossner, 1964, p.151)

4.3.7 The Degree of Benefit

Case law indicates that the social benefit of an act can determine the level of safety that is 'reasonable' even to the point of risking, and losing, one's life. In New York, in 1971, the jury found that a man who was killed saving a child from an oncoming train, had not exposed himself to an unreasonable risk. In this case the probability of harm and the magnitude of the consequences were both high, but the social value of the risk taken made it 'reasonable'. If, instead of a child, the man had saved a dog or his car, the decision would almost certainly have been reversed.

4.3.8 Knowledge of Risk

What is considered 'reasonable' is usually defined in terms of what the parties involved knew, or could be expected to know, in the circumstances. The amount and quality of information in relation to the risk is therefore a key factor in legislation and case law dealing with the criterion of reasonableness. Professional liability law in some jurisdictions distinguishes between patent risks (obvious to an ordinary person) and latent risks (hidden flaws). The degree of warning required by labelling and sign posting is commonly defined by statute for dangerous products and places. Case law tends to confirm the importance of making as much information as possible available about a risk in order to render it a more 'reasonable' one. In common law, the doctrines of negligence and assumption of risk hinge upon defining the degree of information and understanding of the risks that existed in each case.

4.3.9 Changes in Legal Interpretations of Policy

Traditionally case law in countries with English Common Law systems has provided a responsive, after-the-fact mechanism for assigning liability for harm already done, since many cases are brought as a result of accidents and injury. As the hazards we are concerned with have longer latency periods between cause and harmful effects, and as causes become almost impossible to isolate from one another, recourse to present common law concepts of reasonableness and negligence will prove of less and less help. This is because, as used in common law, the criterion of reasonableness is weakened for the purposes of risk assessment by its dependence on *demonstrated* cause and effect.

In legislation, however, reasonableness is still considered to be an effective and flexible basis for making decisions even in relation to long-term genetic risks. In the recent and far-reaching US Toxic Substances Control Act 1976, a new shift of emphasis from past to future proof of reasonableness is shown. Previously the best and only convincing proof was what has happened in the past. The new statute requires only a *reasonable likelihood of future harm*. Future case law may further enshrine or dismantle government agencies' mandate to make judgements on the basis of what they consider to be reasonable or unreasonable risks. The continued widespread use of 'reasonableness' in legislation indicates its effectiveness in public policy risk assessment.

In the US, the role of the courts has extended into a review of federal government agency decisions and in-house procedures with regard to environmental risk assessment. This process was encouraged by one section of the US National Environmental Policy Act 1970 (NEPA) which required each federal agency to prepare a detailed statement of the environmental impact of every major federal action that might significantly affect environmental quality (Section 102(s)C). This provision applied to federal projects ranging from roads and other major construction works, to decisions about pesticide registrations, pollution control or even the impacts of *legislation* itself such as the control of recombinant DNA experiments. Thousands of such impact statements have now been filed (Anderson and Daniels, 1973). The American courts have not only played their traditional role of interpreting the meaning of terms in the legislation such as 'major action', 'federal action' and 'significantly affect', but have used the legislation to take a close look at the adequacy of environmental risk assessment procedures by government agencies. The courts have also required government agencies to explicitly assess the risks and benefits of the decision it has reached, and the *risks and benefits of alternatives* it has considered. Federal risk assessments coming under NEPA are now required to include

— clear statement of the reasoning that supports the decision taken
— elaboration of the risks involved
— discussion of alternatives, and their risks
— public participation and inter-agency review
— established agency procedures for principled decision-making

These changes in US federal risk assessment procedures were to some extent envisaged by the promoters of the bill but the courts showed themselves very willing to use the legislation to provide a judicial review of environmental risk management in the USA. It is a process that may well develop in other countries.

4.4 ECONOMIC CONSIDERATIONS

In many areas of contemporary public policy, economic considerations are paramount. 'Is it worth the cost?' and 'Can we get better value for money in some other way?' are the questions that must be satisfactorily answered. The incorporation of such economic considerations into the assessment of environmental risks has proved to be a difficult and frustrating task. There are three reasons why this is so.

First the methods developed for economic evaluation have not won widespread acceptance in risk assessment. Second, there is often lack of agreement on the objectives to be satisfied, and third, the necessity to place a monetary value on human life, if risks are to be evaluated in economic terms, is a source of deep-seated conflicts of value and belief.

The two main methods in use are cost-benefit analysis and its variant risk-benefit analysis. Beyond these two a series of other methods (Table 4.3) have been undergoing development and have been used in a small number of applications. All the methods listed can be useful in dealing with particular

88

Table 4.3   Methods for Evaluating Risk

Main Objective	Mode of Comparison	
	With benefits	
Utility maximization	Cost-benefit analysis Risk-benefit analysis	} Single dimension
	Multiple-attribute utility theory Social judgement theory Decision analysis	} Multiple dimensions
	With cost	
Risk reduction	Cost-effectiveness	
	With other risks	
Risk rationalizing	Comparison with natural levels Comparison with the risks of alternatives Comparison with other (unrelated) risks	

risk problems, but each of them has its critics, and each is subject to a number of limitations, both practical and theoretical. All of the methods require extensive information which may be costly to acquire. There is also a commonly expressed concern that the methods may give an impression of precision that is not warranted by the procedures used. The methods also require some simplifying initial assumptions to be made which can have a major effect on the outcome but which may not be apparent to the decision-maker. One has only to vary certain assumptions to change the result radically.

Perhaps the most positive statement that can be made about these methods is that they are an aid in *ways of thinking* about risk problems more than a set of tools for providing unambiguous solutions. They can encourage careful and logical thinking. At worst these tools can be a means of self-deception. If a decision-maker is offered a precise answer on the basis of an analysis which relies on doubtful assumptions, then the analysis should be treated with skepticism.

The methods being developed are listed in Table 4.3. All methods involve a consideration of a particular objective and a mode of comparison. In essence the questions asked are 'What objective is being sought?' and 'How does this risk compare with other considerations that are involved in obtaining the objective?'

Three quite different objectives are in use in evaluating risk. These are utility maximization, risk reduction and risk rationalizing. Utility maximization means simply trying to gain the highest net value from any risk management situation. In theory the level of risk is set at that point from which a move in *either* direction, i.e. an increase *or* a decrease in risk would lower the net value or utility. Some would argue that this is in fact the only objective worth considering and that risk reduction or risk rationalizing are not really objectives at all.

The objective of *risk reduction* arises in part because the methods of analysis designed to help achieve utility maximization are deficient in several respects and their results are always subject to question and qualification. To adopt the objective of reducing risks as far and as long as it is cost effective to do so, simplifies the analysis and avoids the problems associated with the methods used to achieve utility maximization. At the same time, of course, it introduces the possibility of new error. In neglecting the question of the ratio of benefits to costs and risks the analysis can lead to a result which would most effectively reduce risk, but beyond the point at which utility ceases to increase.

In order to provide a check on such error as a specified level of risk for one activity can be compared with other existing levels of risk. In the strict sense *risk rationalizing* is not really an objective at all. Nevertheless risk rationalizing comparisons are in common use in public debates concerning acceptable levels of risk.

4.4.1 Cost-Benefit Analysis

The question this method seeks to answer can be stated very simply: do the benefits to be gained from this activity make it worth the cost? Cost-benefit analysis is essentially a way of duplicating in the public sector the private sector analysis of profitability. The purpose is to reflect the costs (one element of which is environmental risks) and benefits of an activity to the society as a whole. A major characteristic of cost-benefit analysis is that the return on an investment is not necessarily realized in terms of a direct cash flow back to the project. A bridge over a river previously impassable in the rainy season, for example, may yield net benefits by reducing the transport costs of all those who use it. These benefits are spread over a large number of bridge users — and through them to the rest of society, because costs are lowered, and greater efficiency is achieved.[8]

Benefit-cost procedures also permit comparison among proposed projects or developments as well as comparison among proposed alternatives for a single project. While problems in estimating the benefits and costs of an activity have been dealt with at length elsewhere (Mishan, 1976) it is important to note some of the major ones to suggest issues which must be confronted in using this method:

(1) In the absence of perfect competition, market prices of goods and inputs are not accurate guides. They have to be corrected, therefore, when used for project evaluation.

(2) There are costs and benefits (e.g. the cost of a human life or the benefit of pleasant scenery) with no *explicit* market equivalents which have to be somehow estimated. The valuation of indirect and 'intangible' costs and benefits is difficult, especially if they involve *irreversible* effects. An obvious example is the loss of 'life and limb'. Various criteria have

been proposed for the determination of the 'price' of a life. A well-known criterion is the use of an index of the expected life-time earnings of the deceased. Benefit-cost calculations rarely involve explicit statements on the price of a life, but each time a decision is made concerning the amount of money it is worthwhile to spend to gain an extra margin of safety an implicit judgement about the value of life is being made (Sinclair, 1972).

(3) Investment in any activity results in costs and benefits *over time*. This requires the estimation of *future* costs and benefits, and their expression in present values. This task necessitates the choice of a social rate of discount or 'time preference'. This is the issue of the relative valuation of present and future welfare. Given the full employment of existing resources, for example, a lower rate of growth of output would imply a higher *immediate,* and a lower *future*, aggregate consumption than a higher growth rate. The question is: How, and by what criterion, do we determine the value of the welfare of the *present generation* relative to the *future generation*. Thus, the present value of the net benefit of a project will be lower the higher the social rate of discount — and vice versa. Intuitively, it is probably evident that the choice of an 'objective' or market-based social discount rate is impossible.

(4) Another issue for which there has been no satisfactory solution is the problem of *distribution*. This question poses itself at two levels:
(a) How can we ensure that the burdens and benefits of a public project are equally shared by members of the public?;
(b) How can we do the same for those who lose or gain *directly* from a public project?
A real solution to the first question is forbidding; that is why it is usually dismissed by the argument that 'eventually' each person's losses and gains will balance given the number and range of public investment projects; *and* that the issues of public investment and redistribution are, in any case, separate policy questions. As for the second, compensations are seldom valued satisfactorily and paid out even less often.

(5) In the absence of perfect competition, market prices are inaccurate guides to real opportunity costs of inputs and outputs. This is particularly so in those countries where economic dualism (a mixture of socialist and capitalist systems), structural rigidities and immobilities, survival of significant barter relationships, and public intervention in domestic and foreign trade are more evident. Therefore, it would be necessary to evaluate costs and benefits by the use of 'shadow prices' (i.e. accounting prices or imported values which more accurately reflect the real cost of goods and resources in the society). In addition, there are some items without explicit market equivalents which have to be estimated.

4.4.2 Risk-Benefit Analysis

Risk-benefit analysis is properly thought of as a subset or particular type of the cost-benefit approach. The focus is on how the risks (as one type of social cost) compare with the benefits. Since the measures of risk cannot readily or justifiably be reduced to monetary terms, the analysis usually takes the form of a comparison of benefits with the injury or fatality rate.

A well-known discussion of this technique is that by Chauncey Starr (1972). Starr estimated the risk of death per person-hour of exposure and compared them with the benefits for a variety of activities. The benefits were calculated by assuming them either as equivalent to the expenditure for the activity or as expenditure plus some additional benefit. The benefits and risks were then plotted against one another, revealing that risks tend to increase by the cube of the benefits and that society accepts much higher levels of risk (by a factor of 10^3) for risks which are 'voluntary' than for those which are 'involuntary'.

A second approach to risk-benefit analysis involves not the extent levels of risk and benefits in society but rather how these are *perceived* by the public. Risks and benefits are both experienced by people and it is useful obviously for the risk manager to know what the public thinks the risks and benefits are. This approach has the added value of building in public concern over the different consequences as well as their likelihood of occurrence. There is some indication that some types of consequences are more feared than others. Cancer, for example, is a particularly dread form of disease and may represent a larger *felt* risk than health or fatality statistics would suggest (Slovic, 1978).

Both of these approaches do provide useful and somewhat different, perspectives. Both also yield information concerning risk-benefit comparisons to the risk assessment process and to the risk manager.

NOTES

1. Level I. Concentration and exposure time at or below which, according to present knowledge, neither direct nor indirect effects (including alteration of reflexes or of adaptive or protective reactions) have been observed.

   Level II. Concentrations and exposure times at and about which there is likely to be irritation of the sensory organs, harmful effects on vegetation, visibility reduction, or other adverse effects on the environment.

   Level III. Concentrations and exposure times at and above which there is likely to be impairment of vital physiological functions or changes that may lead to chronic diseases or shortening of life.

   Level IV. Concentrations and exposure times at and above which there is likely to be acute illness or death in susceptible groups of the population.

   (WHO, 1964)

2. The wording of the principle is 'If a chemical is known or shown to be a carcinogen it will not be permitted to occur as a residue in food'. *Pesticides Safety Precautions Scheme.*

3. Delaney Clause for food additives is contained in Sec. 409(c)(3)(A) of the Food Additives Amendment of 1958. 21 USC Sec. 348(c)(3)(A) (1964). Delaney Clause for food colour is contained in Color Additive Amendments of 1960 Sec. 706(b) (5)(B) 21 USC Sec. 376(b)(5)(B) (1964).

4. The minor difference in wording is the replacement of the total cost of application of technology in relation to effluent reduction benefits to be achieved by such reduction with the cost of achieving such effluent reduction for the best available technology/Sec. 304(b)(2)B/. This difference is pointed out by Cox, Doak, C., 1972, 'The Best Practicable Control of Environmental Discharge', *Proceedings 31st Annual Meeting of Hawaiian Sugar Technologists,* November 1972.

5. In Common Law to succeed in an action of tort (or wrong) the plaintiff must establish (a) what a duty of care is and (b) that the defendant failed to take *reasonable care* in fulfilling that duty. Thus the plaintiff must prove *negligence*. In an action of nuisance, the plaintiff does not need to establish either of these points. He must show only that the nuisance exists and that the defendant is the relevant owner or occupier. It is up to the defendant to try to excuse himself.

6. The UK Alkali Act 1863 enabled secret effluent limits to be set for individual chemical works as well as prohibiting disclosure of actual discharges.

7. This is true even for some of the most far-reaching recent environmental legislation. For example,

    US Federal Environmental Pesticide Control Act 1972 — P.L. 92-516 'Unreasonable adverse effects on the environment. /defined as/ any reasonable risk to man or the environment, taking into account the economic, social, and environmental costs and benefits of the use of any pesticide'. /78/

    US Toxic Substances Control Act 1976 — P.L. 94-469 'If the Administrator finds that there is a *reasonable* basis to conclude that ... such activities present or will present an *unreasonable* risk of injury to health or environment ...'. Sec. 5.

8. The procedure for developing a cost-benefit analysis may be described simply as follows. Let the 'life' of a project be n years. Then benefits accrued and costs incurred in each year may be estimated to be $B_1, B_2 \ldots B_n$; and $C_1, C_2 \ldots C_n$. Therefore, the total *net* benefit of the project would be equal to the sum of the *net* benefit for each one year, i.e.

$$\text{Total net benefit} = (B_1 - C_1) + (B_2 - C_2) + \ldots (B_n - C_n)$$

$$\text{Or, symbolically,} \quad \sum_{t=1}^{t=n} (B_t - C_t)$$

The sum of the benefits minus costs over the life of the project.

But, since — even at constant prices — a dollar's worth of *present* consumption is higher than *future* consumption, the nominal value of future costs and benefits must be reduced to their present worth by discounting them at a social rate of discount, r. Hence:

Worth = Total present net benefit

$$\frac{(B_1 - C_1)}{(1 + r)} + \frac{(B_2 - C_2)}{(1 + r)^2} + \ldots \frac{(B_n - C_n)}{(1 + r)^n}$$

Or, symbolically,

$$\sum_{t = 1}^{t = n} \frac{(B_t - C_t)}{(1 + r)^n}$$

CHAPTER 5

Managing Environmental Risks

Let us imagine a hypothetical figure, the National Chief Environmental Risk Manager. His task is to allocate manpower and technical resources located in various government agencies to most effectively control the risks in his country. What does he need to know to carry out his task?

First, what are the risks with which he should be concerned? Which hazards cause most damage, and when and where do they occur? Second, which government agency is formally responsible for each of the tasks related to managing each hazard — monitoring, standard setting, enforcement etc. We do not need to go any further to point out that few, if any, countries in the world could adequately respond to their Chief Environmental Risk Manager's most basic information needs for allocating risk management resources.

This chapter begins, therefore, with a description of a management strategy that has not, to our knowledge, been completed anywhere on a national scale and to cover all environmental risks (although the process has started in Sweden). This is the development of a national profile of environmental risks which can enable hazards to be ranked into priorities for different types of action. Once this national risk profile is known, the task of matching risk management tasks to government departmental functions can begin. No management executive starts with a clean slate — he inherits agency structures with their traditional jurisdictions, ways of operating, and areas of expertise. New problems however, often demand changes in procedures and organization. These changes are often resisted by those affected and can ultimately turn out to be harmful if they occur too often or too drastically. The job of the Chief Risk Manager would be to steer a careful course between the needs for rational management of complex problems and the needs of administrative structures for continuity and clearly defined tasks.

Administrative arrangements in national and local government vary enormously around the world. This chapter cannot hope to discuss each one of them in the context of environmental problems. Rather, the route chosen here is to indicate some organizational changes which can enable traditional functional departments to cope with interdisciplinary (and thus interdepartmental) problems. Also discussed are different management tasks to show what is involved in each one and the range of different activities that the tasks collectively involve. Finally the different kinds of management issues that are associated with different environmental problems are described. For

95

example, those that arise from many small environmental impacts (such as subsistence farming) and a few large impacts (such as industrial developments), with particular reference to developing situations.

5.1 DEVELOPING A NATIONAL RISK PROFILE

One outcome of a complex government machinery with different departments looking after Health, Agriculture etc. is that information becomes decentralized and scattered. This is particularly true of information about environmental problems, part of which fall under almost every department's area of interest. The net result is that data, even on the statistical incidence of different risks, do not become assembled together. Thus any cross-hazard analysis becomes difficult and the ordering of priorities for action is done without a sound understanding of the relative magnitudes or effects of different problems.

One way to counteract the division of information is to establish a procedure for compiling a national risk profile. As a first step, simple actuarial data on the number and magnitude of different hazards that have occurred in the country can be compiled from any services available — official and private records held in different organizations, newspaper reports, private journals and log books, and even folk records. From such sources a picture can be built up of the
— numbers of events
— their magnitude and effects
— where and when they took place
— who the victims were etc.
Assembling these data over time provides information on trends over time in their various characteristics.

In many countries this kind of exercise will produce as many gaps in knowledge as acceptable data. It will be found that basic information on some risks is simply not known. This is, in itself, useful since the gaps can indicate priorities for (a) information searches, such as research and monitoring, and (b) administrative changes to ensure that aspect of the problem is covered by someone. The orderly arrangement and portrayal of what data are, and are not, available is a valuable first step towards a national data bank for environmental risks.

Since much of these data will be held by different government departments and non-governmental agencies, one route towards collecting them is to ask each agency to set out the risk profiles for their own area of jurisdiction. This will generate sectoral profiles and priorities, such as for workplace risks, agriculture, foodstuffs and industrial processes. From these sectoral profiles, a national data base on risks, together with priorities for action can be developed using interdepartmental committee structures, centralized planning agencies or specialized risk assessment advisory bodies, according to normal government procedures.

Figure 5.1   Data sheet for compiling a risk data base

Environmental risk characteristics / Management capabilities	WHAT IS OUR STATE OF KNOWLEDGE?											WHAT IS OUR MANAGEMENT CAPABILITY?										
	Do we know anything?	Are probability estimates available	Are measurements available? — field	— laboratory	Has it been modelled?	Was this anticipated?	Who knows about this?	Where has there been similar experience	What is likely public/political response?	*What resources do we need to find out/respond?* — emergency response	— legislative authority	— technical/equipment	— manpower	— financial	*What resources do we have?* — emergency response	— legislative authority	— technical/equipment	— manpower	— financial	Which government agencies have jurisdiction and capabilities in these areas?
SOURCE — what are sources/causes																				
— who is responsible																				
— what are initiating events																				
— when do they occur																				
— where do they occur																				
— what events have already happened																				
PATHWAYS — what are environmental pathways																				
— how fast does risk travel/develop																				
— what transformations take place (chemical/physical)																				
— does risk diffuse or remain focussed																				
EFFECTS — what are/will be beneficial effects																				
— what are harmful effects																				
— are effects reversible/irreversible																				
— are effects chronic/acute																				
— are effects teratogenic/mutagenic																				
— what are dose-effect relationships																				
— where are effects felt (geographically)																				
— where are effects felt (demographically)																				
— what are harmful environmental effects																				
— what evidence for synergistic effects																				

One example of a 'knowledge inventory sheet' for describing the state of the art in environmental risk management is given in Figure 5.1. This is suggestive of how the data gathering might be arranged rather than a model that is being used in any specific country. Such a sheet would ideally be filled in for each major hazard within a country. The implications for establishing research programmes, monitoring systems and organizational changes can then be discussed on the basis of such compilations of knowledge.

The task of collating risk data is thus a twofold one:

(1) To describe, for each risk, what is known about it (such as indicated in Figure 5.1)
(2) To develop a 'national risk profile' or list of major risks affecting the country together with priorities for action.

5.2 INSTITUTIONAL ARRANGEMENTS

The organizational structures, both within and between government departments, and the nature of the links between them and the public, play important roles in the risk management process. Both are related to the basic 'style' of risk management and government generally.

Environmental risks are characteristically multidimensional problems which cut across the normal jurisdictions of government departments. Put simply, most government structures are inadequately designed to manage environmental risks. Rarely if ever, are the different technical specialists found within one department that are required to deal with, for example, a pollution risk caused by industry and passing through the air, water and soil to be ingested by plants and animals and eventually through agricultural products to man. More likely, these areas of expertise and administrative jurisdiction fall within several departments such as Labour, Trade and Industry, Environment, Water Resources, Agriculture and Health.

In the UK, for example, the chief risk management authority, the Health and Safety Executive, is linked to Parliament through a somewhat awkward arrangement of three ministers (Employment, Environment, and Industry; though the Employment Secretary is normally the most actively involved). When it comes to important planning decisions involving an element of technological risk, the Environment Secretary is responsible, but for energy related matters, both he and the Energy Secretary will be involved. In practice, however, major decisions involving more than two departments of state will be made by Cabinet Committee or full Cabinet. Matters relating to toxic chemicals are dealt with by the Department of Trade, or Agriculture, Fisheries and Food. In each case a whole series of coalescing advisory bodies are normally involved, all working in close association with the private sector who are often creating the very problems needing regulation. Independent appraisal is coopted also on a confidential basis. So relationships to Parliament are good but controlled, and the opportunities for full independent scrutiny limited (but not absent) while the public is usually kept in the dark.

Table 5.1 summarizes the number of government agencies in developing countries which have responsibilities for specific aspects of the environment. Almost every country listed has several government agencies sharing overall responsibilities for some environmental problems. Natural resources, particularly water, soil, flora and fauna are typically shared between 3 to 5 agencies. Water for example commonly falls under the jurisdictions of Agriculture, Forestry, Irrigation, Public Works, Industry and Rural Development. A few countries, notably Ghana, India, Israel, Ivory Coast, Philippines, South Africa and Thailand have overlapping interests in government agencies to the extent that 10 or more departments may be involved in managing one environmental problem.

Part of the rationale for these multidepartment organizational structures lies in the different tasks that need to be undertaken to manage 'one problem'. Take, for example, the control of pesticides. Table 5.2 illustrates the eleven separate tasks required by different national legislation for countries in the Asian-Pacific region. These include monitoring of the environment and food, licensing of manufacturers, chemical formulae, dealers and applicators, registration of pesticides, analysis and import controls. The number of enforcement agencies ranges from one in Papua New Guinea and Thailand to six in the Republic of China-Taiwan. Figure 5.2 gives examples of two of these national organizational structures for pesticide control, Taiwan and Canada. The Canadian structure is complicated by a parallel set of departments and committees at the regional (provincial) level to those of the federal government.

Where different agencies are involved together several administrative problems may arise:

(1) Uncertainty may exist about exactly which agency should take responsibility so that no action is taken, or it is delayed.
(2) Interdepartmental rivalries and jealousies may result in information being withheld between agencies which needs to be shared in order for the best solutions to be found.
(3) Each agency tends to have its own particular interests and constituency of political and public support so that interagency conflict may ensue, rather than cooperative problem solving.
(4) Technical expertise may be too divided between different agencies to enable any one of them to put together the needed scientific and managerial team.

These organizational issues arise not only at the national level but can be exacerbated by similar cross-jurisdiction problems at regional and local levels. They also occur within political decision-making structures. For example, in the USA the Congress Committee organization can lead to differnt parts of the same legislation being worked on by different committees so that the resulting Acts may not be coherent. Legislation covering pesticides regulation, for example, comes under the concern of the House and Senate Committees of Agriculture, Commerce, Merchant Marine and Fisheries and the Government Operations Committee.

Table 5.1 Numbers of National Government Agencies with Environmental Responsibilities in 63 Developing Countries (data abstracted from Johnson, Johnson and Gour-Tanguay (1977))

Country	General Policy	Air	Water-fresh	Sea Water	Soil	Fauna & Fish	Flora & Forest	Non-renewable resources	Noise	Solid Waste	Hazardous Substances	Land Use Planning	Habitat	Economic Development	Protected Areas	Environ. Modif.	Population	Envir. Education
Afghanistan	1		4		1		1	4									1	
Algeria	10				1	1	1								1	1		
Argentine*															1		2	2
Barbados										1	1	1	1	1	1	1		2
Benin			1	1	1	1	1					1						
Botswana	1		2		1	3	2	1				2			3	1		1
Bulgaria	1		3		1	2	4		1		4		1		8			
Burma			1	9	1		1	1			1		1		3			
Burundi																		
Cameroun			1			2	2								3			
Central Africa	3		4			3	3								3			
Chad	1	1	1												5	1		
China (Taiwan)	3		2		4		2	2				3	2		1		5	7
Congo							5	2							1		2	6
Cyprus	3	4	4	2	3	4	2	1	1	1	4	1	4				1	2
Egypt					1	3					1		1	1			1	1
Ethiopia			1		1	2	2							1				
Gabon			2	2		4	2					10			2			2
Ghana	1		9			3	2	3	1	3	7		1		4	3	1	2
India	2	9		2	9	3	1		1		3		2		1	1		3
Indonesia		1	1	1	5	2	4					1			6			3
Iran			2	1	3	3	2	2			1		1		3		1	1
Iraq	1	3		2	2	1	1									1	1	4
Israel	1	2	4	5	1	10	5	10	2	2		16	4	3	5	3	1	4
Ivory Coast	2	3	8	10	2	2	9	4	3	3		1	5	4	5	3	1	5
Jamaica	2	2	6	4	2			1				1	1		3		2	

Country	1	2	3	4	5	6	7	8	9	10	11	12	13	14	15	16	17	18	19
Kenya	1	1	1		2	7	1	1			4	1	2	1	1	1	3	2	3
Korea (South)	12					2					3				5				
Kuwait								1				1		1					1
Liberia	1		2	1	4		5	3		1		1	2		2			1	5
Lybia					1	2						1							
Malawi	1	3	1		1	3	1					3	1	1	2	2		1	2
Malaysia	4				1	8	6				1		7	1	4			2	4
Mali	1		2		1	1	2								1				
Mauritania			1	1		1	1	1							1				
Morocco			1		3	5	4				1			1	2			1	1
Nepal	1					1	2						1						
Niger	3		4		4	3	1				1				2				
Nigeria	2	1	2			2	2					1			1			1	2
Pakistan			3		1	3	2	6			1				3			4	1
Philippines	5	3	4	2	1	4	6	1	3	2	1	2	2	2	3	2	4	3	
Qatar			2	2	1		2	2	1				4		2				1
Salvador		3				1	2												
Saudi Arabia			1		1	1	1	1							1				1
Senegal		2	3	3	1	5	3									6			
Sierra Leone			1	1	1	4	2	4				1	1					1	
Singapore	1	1	2	1		1	1	1		1	2	2	2		4		1	2	
Somalia	1		1			3	2	1											1
South Africa	6	12	9	11	8	15	7	5	8	1	4	11	6	1	21			3	10
Sri Lanka	2		2	3	2	5	5								8			1	1
Sudan			3		3	4	2								4				3
Swaziland			1					1							1				
Tanzania			3		2	5	3	1				3	3	2	5			1	4
Thailand	2	1	2	1	3	7	3	3	2	2	5	4	3	3	2			4	4
Togo		3	3	5	1	5	3	3	3			1		3	7				1
Trinidad & Tobago		1	1	4	2	2	2	1	1	1	1	1	1	1	3			2	
Tunisia	1		1		1	2	1	2				1	1		2			1	3
Uganda			1	4	2						2			1	2			1	1
Upper Volta			3	3	1	3	4	3		2	1	4		3	3			2	2
Yugoslavia	1	2	4	4				1	3	3	3	1	1		4			1	1
Zaire	2	2	5	2	4	5	5	4	2	3	2	5	5	5	3	2		4	3
Zambia	1			2	4							1			1			1	2

Key: 3 — number of government agencies responsible (including major departments within agencies); blank — no agencies reported
* data for freshwater in Argentine missing

Figure 5.2 Organisational structures for pesticide control in the Republic of China-Taiwan and Canada

A. ORGANISATIONAL STRUCTURE FOR PESTICIDE CONTROL IN REPUBLIC OF CHINA-TAIWAN

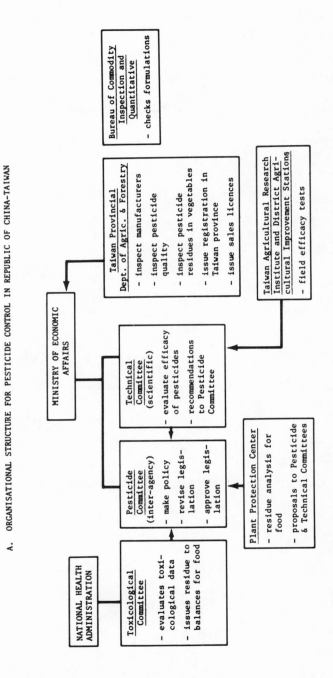

B. CANADIAN ORGANISATIONAL STRUCTURE FOR PESTICIDE CONTROL

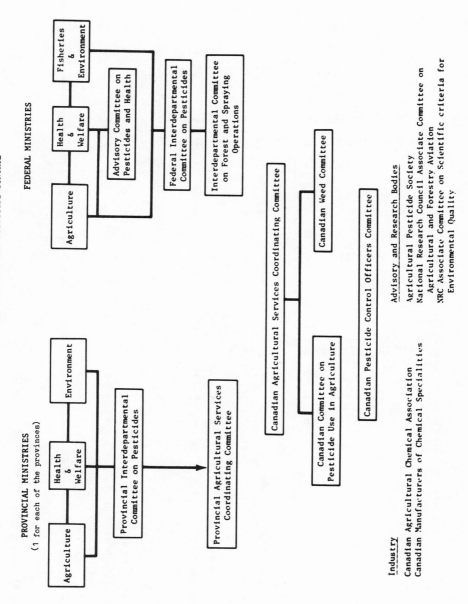

Table 5.2 Pesticide Legislation in Countries in the Asian-Pacific Region

Present legislation requiring:	Australia	Canada	China-Taiwan	India	Indonesia	Iran	Japan	Korea	Malaysia	New Caledonia	New Zealand	Niue	Pakistan	Papua New Guinea	Philippines	Thailand	Tongo	USA	Guam	Total no. countries surveyed having legislation
Registration of pesticides	x	x	x	x	x	x	x	x	x	x	x	o	x	x	x	x	x	x	o	17
Guaranteed analysis of pesticides	x	x	x	x	x	x	x	x	x	o	x	o	x	x	x	x	o^1	x	o	15
Import controls	x^2	x	x	x	x	x	x	x	x	x	x	o	x	x	o	x	o^1	x	x	15
Licensing of manufacturers	x^2	x	x	x	x	o	x	x	x	o	o	o	x	x	x	x	o^1	x	o	13
Licensing of formulations	x	x	x	x	x	x	x	x	x	o	o	o	x	x	x	x	o^1	x	o	14
Licensing of dealers	x^2	x	x	x	o^1	o	x	x	x	x	o	o	o	x^1	x	x	o^1	x	x	14
Certif. applicators	x^2	x	o	o	o^1	o	o	o	x	o	x	o	o	x	o	o	o^1	x	x	10
Mandated uses	x	x	x^1	x	x	x	x	x	o	x	o	o	o	o^1	o	o	o^1	o	o	9
Residue tolerances in food	x	x	x	o	o	x	x	o	x	o	x	o	o	x^1	x	o	o^1	x	o	10
Monitoring of foods	x	x	x	o	o	x	x	x	x	x	x	o	o	o^1	o	x	o^1	x	x	12
Monitoring of environment	x	x	x	o	o	o	o	o	x	o	x	o	o	o^1	o	o	o^1	x	x	8
No. of agencies involved in enforcement of legislation	3	?	6	2	?	3	2	2	3	5	4	0	2	1	3	1	?	4	2	

Key: x = yes;   o = no
Footnotes: 1 — Regulation pending
               2 — Only in some states

Source: Mootooka, 1977.

There are several ways to try to mitigate these problems; the creation of large 'super-agencies'; the improvement of coordination between departments; the transformation of departments from purely functional structures to regional responsibilities and the development of what are called 'matrix organizations'.

5.2.1 Super-agencies

In recognition of the administrative problems that arise when several agencies have responsibility, some countries including Kenya and Thailand have established departments with special responsibilities for the environment. In the USA the Environmental Protection Agency has an explicit charge to look after the interests of the environment — interests which are often overridden by the economic development concerns of other agencies concerned with Agriculture and Industry, for example. Inevitably, agencies with jurisdiction over the range of environmental problems that exist in any country, become very large. The UK Department of the Environment has been described as a 'super-department' which is concerned with managing everything from pictures in historic buildings to pollution in open spaces.

The creation of such large departments produces problems of their own. Special information units need to be set up within them to communicate to other sections what each section is doing. The massing together of people under one name does not necessarily solve either communication problems nor intersectional rivalry. Nor does the existence of a large department necessarily mitigate the cross-jurisdictional problems. A large Department of the Environment must still cooperate with departments representing health, industry, labour and agriculture if problems of industrial pollution or agricultural pesticides affect human health and the environment. However, the value of an agency such as the US Environmental Protection Agency lies more in its special concern to look after the commonly neglected interests of the environment than in its large organization.

In many developing countries, jurisdiction over environmental concerns does still lie largely in one or two leading departments. Agricultural and Rural Development agencies have wide and often sole powers in many countries over natural resources and their economic development, whereas Public Health departments usually share responsibility (where they have it) with other agencies. Planning Departments, where they exist, also seem to have wide powers over the environment, and have the advantage of a centralized planning function which may also include coordination between other agencies.

5.2.2 Coordination between agencies

Given the fact that risk management is becoming a more and more comprehensive and interventionist process, the question of coordination of its functions with existing agencies operating in related areas is important. In the past this linkage has often been *ad hoc* depending largely on the personalities and experience of the responsible officers, on arrangements developed after accidental events, and in response to proposals made by commissions of inquiry from time to time. It is now apparent that coordination between risk management agencies needs to be more comprehensive and consistently developed.

Already there are signs in terms of guiding legislature principles that this is taking place in some countries. For example, in the arena of pollution abatement and the control of toxic materials, considerable progress has been made in

(1) Coordinating the work of agencies responsible for environmental quality within the workplace with those outside the workplace;
(2) Relating the work of air pollution control authorities to those responsible for water quality; and
(3) Coordinating the activities of international organizations with respect to standardization of environmental quality, monitoring and reducing transnational flows of polluting substances.

In the arena of planning development, the growth of environmental impact analysis, first in the USA, and subsequently in most western countries in some form or another has led to a much more comprehensive planning function, complete with advisory committees, panels of adjudication and public hearings which often require agencies to show that they have responded to each other's initiatives.

However, in most instances, the potential for coordination is much greater than the practice, so it is necessary to ask what impediments impair interagency coordination in risk management and how they can be overcome. Two points emerge:

(1) Tradition and custom backed by legal guidelines often isolate agency responsibilities. For example, pollution control and planning authorities in the UK must consult, but need not listen to each other.
(2) The personality and experience of the responsible officers. This is obviously a delicate matter to investigate, but is certainly very pertinent. Informal consultative arrangements working on the basis of trust and respect is a vital aspect of good risk management, and cannot be legislated. Despite its potential sensitivity, this is a most important area for task evaluation, because risk managers do pride themselves in their professional responsibilities and informal consultative arrangements.

5.2.3 Organizational alternatives to functional structure

The structure of an administration is a large factor in its capacity to recognize and deal with interdisciplinary problems, whatever the individual capabilities of the people working within it. Conventionally, government departments have a sectoral structure in which responsibilities are divided functionally as opposed to regionally. This is true for most industrialized and developing countries. Thus in any one part of a country, departments of social security, health, water, agriculture, trade and industry, will all have a part of the administrative pie. They each receive their operating budgets from a central Treasury and to some extent are in direct competition for financial, technical and manpower resources as well as political support. This typical arrangement is least well adapted to environmental management and has been termed the 'administrative trap' (Baker, 1976).

To take the example of rangeland management, which has been a key problem in the Saharan drought areas: the sectoral structures of most of the African governments involved led to different departments developing strategies for water, animals, marketing and livestock health often quite independently. Key departments for rangeland management for some African countries are given in Table 5.3. The results were that water was provided in some areas without plans for controlling either grazing or population influx, and livestock patterns were changed without adequate marketing arrangements. In Uganda, for example, Animal Industry and Agriculture are two separate administrations so that crucial links between them in developing policies for semi-pastoral and agricultural tribes such as the Karamojong are difficult to achieve. When problems are perceived within one sector, projects are started which tended to patch up the symptoms where they appeared rather than considering the 'problem region' as a spatial set of interconnected symptoms. These are not the fault of the individual departments who had no power to act outside their limited jurisdictions but a weakness of the overall government structure (Baker, 1976).

Again, these structural deficiencies for managing environmental problems at the national level are often repeated *within* departments in the way their divisions relate to one another, and they are exacerbated by the similar sectoral structures of any international and bilateral aid agencies. One solution is to introduce a regional planning and coordination function between the national departments and their field stations or projects in the regions (Figure 5.3).

The advantage of this arrangement is that there is a chance for individual projects to be considered in an interdisciplinary manner. Also, integrated and more flexible (less 'blanket') policies have a chance of surviving but the individual national departments still retain their autonomy and authority. The career structure of their personnel is preserved and they are more likely to favour the change in structure. In particular, no one ministry is given pre-emptive power over others. The disadvantages are that the regional planning and coordinational level may not be sufficiently influential to force sectoral plans to be modified, and that the demand for manpower and other resources to administer government policies is likely to be increased. Some countries are moving towards a regional approach to planning. These include Guinea, Malagasy, Niger, Mali and Pakistan who are all trying to integrate rangeland management through regional administrative structures.

Another way to approach the problems of functional organizations is to develop a 'matrix' structure (Figure 5.4). This simply means that alongside the normal functional divisions are established interdisciplinary or inter-

Table 5.3 Administrative Structure of Selected African Countries for Rangeland Management

Country	Ministry most closely charged with the management of rangeland
Botswana	Ministry of Agriculture
Chad	Ministry of Agriculture and Stockbreeding
Ethiopa	Ministry of Agriculture
Guinea	Ministers for Local Development for Regions: Ministers of Rural Development
Kenya	Ministry of Agriculture and Animal Husbandry
Malagasy Republic	Ministry of Territorial Planning Ministry of Rural Development
Mali	Ministry of Production
Mauritania	Ministry of Rural Development
Niger	Ministry of Saharan and Nomadic Affairs
Rwanda	Ministry of Agriculture and Livestock
Senegal	Ministry of Rural Development
Somalia	Ministry of Rural Development and Livestock
Sudan	Ministry of Agriculture
Uganda	Ministry of Animal Industry
Upper Volta	Ministry of Agriculture, Cattle Breeding, Rivers, Forests and Tourism

Source: Baker, 1976, p.250. (Reproduced by permission of Edward Goldsmith, publishers)

departmental special teams or projects. These draw on the functional divisions for manpower and technical back-up to put together a group of people with the different skills needed for a particular project. These people go back to their functional divisions when the project is completed or when they have served on it for an agreed period of time.

The advantage of a matrix organization is that it can accomodate the needs of special interdisciplinary problems without breaking down the functional structure. The project structure is varying as problems are solved and new ones emerge so that it is not encumbered with a static set of manpower, but can develop teams specially put together for eachproblem. For people working in matrix organizations, the chance to work in a challenging interdisciplinary project is usually attractive and stimulating whilst their permanent 'home' in a functional department, usually with others of similar training (e.g. engineering, medical), gives them career stability and the needed association with members of their own professions. Matrix organizations are being used with success by many large private companies in western industrialized countries and are now being tried by some government departments in North America (Davis and Lawrence (1977)).

a) Sectoral
   (by function)

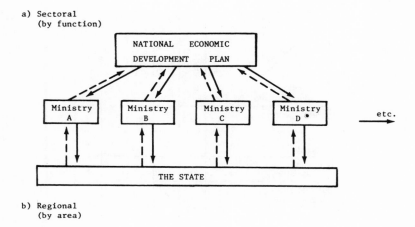

b) Regional
   (by area)

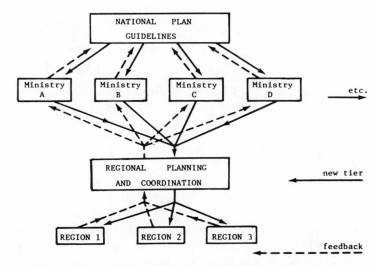

Source: Baker, 1976, p.250. (Reproduced by permission of Edward Goldsmith, publishers)

Figure 5.3 Functional versus regional organizational structures

5.3 RISK MANAGEMENT TASKS

The risk management process can involve a range of different tasks, or management control options. Some of these are carried out largely within government departments, while others are located within the political or public sectors. Not all of the management functions are necessarily applicable to every kind of environmental risk. Major areas of management control are research and monitoring, the drafting of legislation and regulations; standard setting; inspection and enforcement; and continuing review of risk levels and the management process itself.

Interdisciplinary projects / Funtional departments	Water supply	Rangeland management	Small scale cooperative ventures	Air pollution from industry	Floods in urban areas	Development of national risk profile
Public health and welfare	X	X		X	X	X
Agriculture and forestry		X	X			X
Animal husbandry	X	X				X
Trade and industry		X	X	X	X	X
Energy			X	X		X
Labour			X	X		X
Natural resources environment	X	X	X	X	X	X
Housing, urban affairs				X	X	X

Figure 5.4 Schematic matrix organization for environmental management based on functional government departments

5.3.1 Research and monitoring

Scientific knowledge about the nature of the risks is the basis for risk management decisions although at times those decisions have to be made in the face of inadequate knowledge. The gathering of scientific data is, in many countries, a task shared between government agencies, universities, private industries, public interest groups, and members of the public. In many countries local people are an as yet underutilized source of environmental information. For many industrial processes, private industry is able to obtain, and pay for, much more information than government scientists can gather. In some western countries much of these data have remained confidential.

Today two trends are emerging: first, governments are undertaking much more research themselves (at greatly increased direct financial cost to the public) and second, private companies are being forced to give more detail about their own research findings to governments in order to have their products registered for sale and use.

The main ways in which research on environmental tasks is conducted are environmental monitoring, health surveillance, laboratory and field experimentation, testing and screening, accident analysis and modelling. These have been described more fully in Chapter 3 and will only be defined and commented on here in the context of administering them.

Environmental monitoring This involves repetitive observations over time from a network of stations which can be compared between stations and between observation times.

Monitoring is a far more difficult and expensive business than is commonly imagined. In many countries, one solution is to essentially let the risk producers (often industry) monitor themselves. The advantage to public authorities of this arrangement is that the polluter bears the costs of monitoring and in any case has the best access to information and to remedial action. The disadvantages are that the system relies on the honesty and public spiritedness of the polluter (even where it runs counter to his own interests). Government inspectorates are thus often acting in the role of back-up monitoring and do periodic checking rather than a comprehensive monitoring programme.

These kinds of arrangements rely heavily on trust between the regulator, the regulated and the public. In many countries, this trust is breaking down as the public learns more and more instances of ineffective regulation and unacceptably high risk levels. There is correspondingly an increased public demand for monitoring to be carried out by independent or public agencies who have no conflict of interests in seeing regulations enforced. The cost of effective monitoring when wholly undertaken by government can become a major demand on national, financial and manpower resources.

In some circumstances, monitoring can be undertaken by the public, especially for rural areas. Accidents (e.g. spillages) are best monitored (reported) by those on the spot rather than setting up an elaborate official surveillance network. River pollution has been monitored by the public (especially fishermen) in the UK who report indicators such as dead fish, smells and foam or coloured discharges. Earthquakes have been successfully monitored in China by the public. These monitoring systems rely on public education about the indicators of high risk and an effective communication system between the public and responsible government officials.

A second set of issues relating to the monitoring task apply to its comprehensiveness, accuracy, and cost effectiveness. Because of the enormous costs involved in monitoring, the cost effectiveness approach requires most urgent attention especially in relation to the accuracy of recording equipment, the spatial and temporal characteristics of the record, and the standardization of the final results to permit international comparison. For example, the European Commission is currently running into some difficulties in trying to get its member states to accept a commonly agreeable monitoring programme for environmental pollutants. The British government is opposed to the existing proposals on the grounds of needless cost. While some risk areas may be over monitored, they claim, others may escape proper investigation. Examples of the latter include the hindsight investigation of environmental impact assessment once major planning developments have been completed, and the full scale assessment of the medical and economic consequences of measures to relieve deprivation, particularly in developing situations.

The special monitoring problems that may occur in developing countries can be subdivided into two classes: scientific and institutional.

(1) Scientific problems. Most of the available information on environmental problems relates to the temperate zones, and it is often dangerous to extrapolate to the tropics where the climate and vegetation patterns are quite different. The associated monitoring systems may then be less than satisfactory. In fact, there is a great need for dose-response experiments in the tropics, leading to realistic sets of environmental criteria, and to guidelines for the design of monitoring systems.

(2) Institutional problems. In many developed and developing countries, certain monitoring programmes have been initiated and managed on an isolated, *ad hoc* and sectoral basis, to serve quite specific purposes. There has often developed a rather loose and sometimes incoherent system of people and organisations sampling, analysing data, and carrying out assessments. Thus the quality of the environmental management systems and the monitoring programmes which provide the data is not only limited by the lack of scientific and technical capability, but also by the organisation of the systems. The latter are strongly affected by the legal, economic, social and political frameworks, and these are evolving rapidly in many developing countries. This can make the organisation of environmental management very difficult. An additional complication is the shortage of skilled manpower to design and implement the desired management structure.

Health surveillance This is the collation and interpretation of health data from monitoring and census services etc. in order to detect changes in the health status of populations. It has been most advanced where hospital and clinical visits are recorded and centralized in a data bank, so that the information they contain is accessible to computerized monitoring and research programmes.

Testing and screening These involve controlled, often standardized procedures for measuring risk sources, pathways and effects, and can be undertaken in laboratory or field conditions. Many tests for the effects of pollutants and drugs on human health are now costly in terms of money, time and technical manpower so that national governments are increasingly being forced to rely either on research by the industries that are promoting the substances, or the results of other governments' experience.

Research into environmental risks is also conducted through *modelling* (Chapter 3) and *accident analysis* which is an after-the-event inquiry into what happened and why. Accidents provide situations that cannot be ethically produced intentionally in experiments, as well as revealing interconnected causes that may have low probability characteristics or be entirely unexpected. Government research capability should include the ability to merit a scientific team to investigate accidents immediately they occur, since some aspects of risk can only be studied in these situations.

5.3.2 Legislation

Legislation relating to environmental risks have been placed on the statute books of most countries in the world. Table 5.4 shows the present areas of legislation for different aspects of the environment in developing countries. Most countries have legislation protecting their animal and plant resources and their fresh water sources. Other well legislated areas include protected areas (national parks etc.), non-renewable resources, soil, and hazardous substances. Environmental areas for which few developing countries have legislation include environmental modification, population policies, solid waste disposal, noise, and air quality.

Although the passing of legislation is a political process, in many countries environmental statutes and regulations are often drafted initially by technical and legal experts within government departments. In countries where several statutes have followed one another to deal with a particular problem, two evolutionary trends can be seen. These are, greater comprehensiveness and an increasingly creative and anticipatory role in environmental management on the part of governments.

For example, pollution control in European and North American countries is evolving from legislation which controlled emissions of particular pollutants at specific locations (e.g. chimney stacks or river outflow pipes) on a case by case basis, through control on a class by class basis, to ambient air and water quality standards which themselves determine what emission concentrations are allowable (Figure 5.5). The legislative framework has moved from a responsive role which facilitated particular decisions to a guiding role for framing pollution decisions within a wider context of social and economic development. Some legislation, notably the US Federal Resource Conservation and Recovery Act of 1976, has developed the comprehensive trend to the point of considering the impacts of activities and control in one environmental medium, such as air, land or water on the quality of the others.

Thus risk-management legislation has become both more *specific* with clearly defined codes of practice and regulations about operation, monitoring and enforcement, and more *comprehensive* in the sense that it now covers:
— occupational risk environments both inside and outside the work place;
— national and international rules and regulations regarding the discharge and distribution of toxic substances;
— the acceptance of planning and other behavioural controls to reduce the impact of environmental damage and natural hazard; and
— the formation of extensive scrutinizing devices to appraise, review and quantify risks in relation to associated benefits both to existing, and to future generations.

Although the point at which this evolution has reached varies tremendously depending on the type of risk and from country to country, there seems every reason to believe that legislation will continue to follow the pathways described: that is, it will become more comprehensive, more specific with respect to standards, monitoring and enforcement, and more anticipatory with respect to potential risks.

Table 5.4 Existing National Environmental Legislation in 63 Developing Countries (data abstracted from Johnson, Johnson and Gour-Tanguay, 1977)

	General Policy	Air	Water-fresh	Sea Water	Soil	Faune & Fish	Flora & Forest	Non-renewable resources	Noise	Solid Waste	Hazardous Substances	Land Use Planning	Habitat	Economic Development	Protected Areas	Environ. Modif.	Population	Envir. Education
Afghanistan																		
Algeria	x		x	x	x	x	x	x				x	x	x	x	x		
Argentine*	x	x	x	x				x							x			
Barbados												x						
Benin			x			x	x								x			
Botswana			x			x	x	x	x	x		x	x		x			
Bulgaria		x	x		x	x	x	x	x	x	x		x		x		x	
Burma			x			x	x								x		x	
Burundi						x	x	x							x		x	
Cameroun																		
Central Africa	x		x		x	x	x	x				x	x		x			
Chad	x		x			x	x				x				x			
China (Taiwan)		x	x		x			x	x	x	x	x	x	x	x	x	x	
Congo	x			x		x	x	x		x	x							
Cyprus	x	x	x	x	x	x	x	x	x	x	x	x	x	x	x			x
Egypt			x			x	x	x			x							
Ethiopa			x	x	x	x	x											x
Gabon			x	x		x	x	x			x				x			
Ghana	x		x	x	x	x	x	x	x	x	x	x	x	x	x			x
India		x	x	x	x	x	x				x		x		x			
Indonesia		x			x										x	x		
Iran			x			x	x	x							x			
Iraq	x		x			x	x	x				x						
Israel		x	x	x	x	x	x	x	x	x	x	x	x	x	x		x	x
Ivory Coast		x	x	x	x	x	x	x				x	x		x			
Jamaica		x	x		x	x	x	x				x	x		x			
Jordan		x	x			x	x	x			x		x				x	
Kenya		x	x	x	x	x	x			x	x	x	x		x		x	x
Korea (South)	x														x			
Kuwait				x				x										
Liberia			x	x		x	x	x				x	x		x			
Lybia	x		x	x	x	x	x	x				x	x					
Malawi		x	x		x	x	x				x	x	x		x			
Malaysia	x	x	x	x	x	x	x	x	x	x	x			x	x		x	x

	General Policy	Air	Water-fresh	Sea Water	Soil	Faune & Fish	Flora & Forest	Non-renewable resources	Noise	Solid Waste	Hazardous Substances	Land Use Planning	Habitat	Economic Development	Protected Areas	Environ. Modif.	Population	Envir. Education
Mali		x			x	x	x	x							x			
Mauritania			x	x		x	x	x			x				x			
Morocco		x	x	x	x	x	x	x			x	x	x		x			
Nepal						x												
Niger	x	x				x	x	x			x	x	x		x			
Nigeria			x	x		x	x	x	x	x	x				x			
Pakistan			x		x	x	x				x	x			x			
Philippines	x	x	x	x	x	x	x	x	x	x	x	x	x	x	x	x	x	x
Qatar																		
Salvador		x	x		x	x	x	x	x	x	x	x						
Saudi Arabia						x						x			x			
Senegal	x	x				x	x	x							x			
Sierra Leone			x	x	x	x	x	x			x							
Singapore	x	x	x	x	x	x			x	x					x			
Somalia		x				x	x											
South Africa	x	x	x	x	x	x	x	x			x	x	x		x		x	x
Sri Lanka			x	x	x	x	x				x	x			x			
Sudan			x	x	x	x			x						x			
Swaziland			x		x	x	x	x			x	x			x			
Tanzania			x		x	x	x	x			x	x	x		x			
Thailand	x	x				x						x			x			
Togo			x		x	x						x			x			
Trinidad & Tobago		x	x			x	x		x		x	x						
Tunisia			x	x	x	x	x	x			x	x			x			
Uganda			x		x	x	x	x			x				x			
Upper Volta	x	x				x	x				x	x			x			
Yugoslavia		x	x	x	x	x	x	x	x	x	x	x		x	x			
Zaire	x	x	x			x	x	x	x	x	x	x			x			
Zambia	x	x			x	x	x	x			x	x			x			
TOTAL	20	19	48	30	36	55	50	35	17	13	34	33	25	7	46	2	9	8

Key: x — legislation; blank — no legislation;
*data for freshwater in Argentine missing

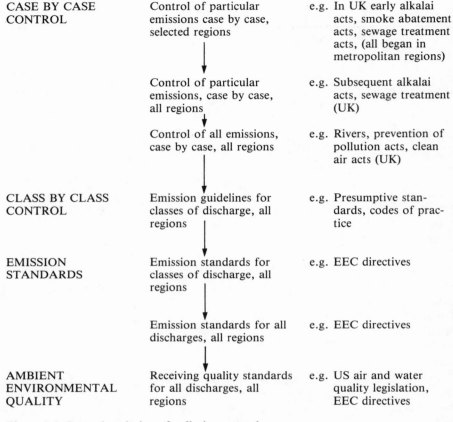

CASE BY CASE CONTROL	Control of particular emissions case by case, selected regions	e.g. In UK early alkalai acts, smoke abatement acts, sewage treatment acts, (all began in metropolitan regions)
	Control of particular emissions, case by case, all regions	e.g. Subsequent alkalai acts, sewage treatment (UK)
	Control of all emissions, case by case, all regions	e.g. Rivers, prevention of pollution acts, clean air acts (UK)
CLASS BY CLASS CONTROL	Emission guidelines for classes of discharge, all regions	e.g. Presumptive standards, codes of practice
EMISSION STANDARDS	Emission standards for classes of discharge, all regions	e.g. EEC directives
	Emission standards for all discharges, all regions	e.g. EEC directives
AMBIENT ENVIRONMENTAL QUALITY	Receiving quality standards for all discharges, all regions	e.g. US air and water quality legislation, EEC directives

Figure 5.5 General evolution of pollution control

5.3.3 Standard setting

Standards Standards are prescribed levels, quantities or values, which are regarded as authoritative measures of what is a safe enough, or acceptable, amount of pollution, contamination or exposure to risk. Standards are usually arrived at in the context of *criteria* which describe the known relationships between risk levels and other factors (see Chapter 4). Standards may refer to (Lowrance (1976)):

Human exposure to risk (e.g. radiation exposure standards)
Effluent standards (e.g. industrial toxic wastes)
Ambient environmental quality (e.g. drinking water quality standards)
Occupational conditions (e.g. length of working hours)
Product, technology or technical process design (e.g. consumer or industrial machines)
Product composition (e.g. processed food standards)
Product or technology performance (e.g. building structures)

Product labelling and advertising (e.g. pesticides)

Product packaging (e.g. pressurized gas containers, child-proof drugs)

Standards vary in the degree to which they are qualitatively or quantitatively expressed and how closely specified or definitive they are. They can, for example, be expressed as a fixed concentration of a pollutant per volume of air, (water or discharge, etc.) above which level concentrations are considered unacceptable, and below which, they are acceptable. A standard can be a single numerical value, or a range of values on the one hand; all the way to expressions endorsing a 'best practicable means approach'. The way in which they can be enforced also varies — it is easier to define when a standard has been broken or executed where it is written down in very specific terms. Partly for this reason, labour groups and public interest groups are often pushing for governments to come up with clearer, more 'black and white' standards, whereas industry generally favours more discretionary terms.

One process which has gone along with more numerical standards is that of also prescribing how they are to be attained in terms of procedures (codes of practice) or equipment (technical codes). This trend of codifying regulations is occurring in the United States and has the disadvantage of implicating the regulatory body as partly responsible for any damage which may ensue where, for example, prescribed regulations have been followed but harmful effects can be shown to have resulted.

Guidelines Instead of legislated standards, which usually have the power of legal enforcement behind them, governments can adopt *guidelines* or recommended standards (presumptive) which specify target levels or desirable standards rather than hard and fast rules and prescriptions. There can be very different rationales behind the option of guidelines rather than specific standards. These include (Doern, (1977)):

(1) Scientific uncertainty about what is an adequate standard;
(2) A concern that set standards are not flexible and easy to change, particularly where technology or products are rapidly developing, so that a fixed standard may prevent improvement;
(3) An unwillingness to enforce tough standards that will be unpopular with industry, the public or politicians;
(4) An awareness that the enforcement resources are lacking and a reluctance to have a 'meaningless' (unenforceable) standard.

Critics of a guideline approach to regulation argue that guidelines will encourage better safety levels only if they can be enforced; or inspectors can apply some leverage to see they are complied with or improved upon. Generally guidelines lack the 'teeth' of legislated standards but in areas where standards cannot yet be determined, it is probably better to have guidelines than nothing.

Some guidelines are in the form of recommendations by prestigious scientific bodies, often international ones, such as the International Commission on Radiological Protection or the World Health Organization

Table 5.5 Comparative National Drinking Water Standards for Selected Countries

Parameter		UNITS	1. International WHO acceptable	2. International WHO allowable	3. European WHO	4. United States	5. Sweden	6. France	7. Bulgaria	8. Tanzania	9. Japan (1968)	10. India (1973)	11. India recommended 1975	12. Israel 1974
Radioactivity	a	pCi/L		3	3-10								3	3
	b	pCi/L		30	80-100								30	
pH			7.0-8.5	6.5-9.2			6.0-8.0		6.5-8.5	6.5-9.2*		6.5-9.2	6.3-9.2	6.5-9.5
Total hardness	as (CaCO$_3$)	mg/L			500			300	450	600*	<300	600	600	
Chlorides	(as Cl)	mg/L	200	600	600	250	25/250	250	250	800*	<200	1000	1000	600
Flouride	(as F)	mg/L		1.5	0.7	0.8-1.7	1.5		0.7-1.0	8.0*	<0.8	2.0	1.5	1.4-1.7
Nitrate	(as NO$_3$)	mg/L		30.0	50/100	45	30	44	30	100	45	50	45	90
Copper	(as Cu)	mg/L	1.0	1.5	0.05	1.0	0.05/1.0	0.2	0.2	3.0*	<10	3.0	1.5	1.4
Iron	(as Fe)	mg/L	0.3	1.0	1.0	0.3	0.2	0.2	0.2	1.0*	<0.3	1.0	1.0	1.0
Manganese	(as Mn)	mg/L	0.1	0.5	0.05	0.05	0.05	0.1	0.1	0.5*	<0.3	0.5	0.5	
Zinc	(as Zn)	mg/L	5.0	15.0	5.0	5.0	0.3/5.0	3	3	15.0*	<1.0	15.0	15.0	15.0
Magnesium	(as Mg)	mg/L	50	150	125			125	50			150	150	150
Sulfate	(as SO$_4$)	mg/L	200	400	250	250	25/250	250	250	600*		400	400	400
Phenolic compounds		mg/L	1	2	1	1			1	2	5	2	2	2
Color	(p.c. scale)	mgPt/l	5	50		15	10		15	50*	<5	25		50
Turbidity		mgSiO$_2$	5	25		3	weak		30cm/y	30*	<2	25		25
Taste														
Calcium		mg/L	75	200					150					
Odor														
Arsenic	(as As)	µg/L		50	50	50	10/50		50	50	<50	200	50	50
Cadmium	(as Cd)	µg/L		10	10	10	10		50	50	<10		10	10
Cyanide	(as Cn)	µg/L		200	50	10	10/20		10	200	0	10	50	50
Lead	(as Pb)	µg/L		50	100	50	20/50		100	100	<100	100	100	50
Mercury	(as Hg)	µg/L					1/5				0		1	10
Selenium	(as Se)	µg/L		10	10	10	10/50		50	50		50	10	10
Polycyclic A.H.		µg/L											200	
Chromium		µg/L		50	50	50	20		50	50	<50	50	50	50
Beryllium		µg/L												
Molybdenum		µg/L												
Strontium		µg/L												
Barium		µg/L		1000	1000	1000			1000	1000				1000

	Blank indicates that data are not available or have not been located.
50/100	The lower value refers to permissible concentrations in water purified by chemical flocculation and slow filtering. The higher value is the permissible concentration in waters much more extensively treated before use.
*	Indicates tentative figures.
<	Indicates 'less than'
ppm	Indicates 'parts per million'

13. Poland	14. U.S.S.R. (1961)	15. U.S.S.R. (1973)	16. Germany (1975)	17. Australia	18. Cairo	19. European Council 1975	20. Greece	21. Korea	22. Philippines	23. Thailand	24. Mexico	25. Czechoslovakia 1963	26. Canada 1968 acceptable	27. Canada 1968 allowable	Range of Standards
				3			3		3		3				3-10
				30			30				10				10-100
9.0		6.5-8.5		6.5-9.0	6.0-8.5		7.0-8.5	5.8-8.0	7.0-8.5			6.5-8.5	6.5-8.3	6.5-8.3	6-9.5
E/L		10ml/L		30/500	4mE/L		100/500	<300							300-600
5		300		200/600	400	200	350	150ppm	200	250			<250	250	200-1000
0		0.7-1.5	1.5	1.5	1.5	1.5	1.5	1ppm							0.7-2.0
		45	90				50						<10	10	10-100
	0.1	1.0		1.0	1.5	1.5	1.0	1ppm	1.0	1.0/3.0	1.0	0.1	<0.01	1.0	<0.01-10
2		0.3		0.3	1.0	0.3	.1	0.3	0.3	0.5			<0.05	0.3	<0.05-1.0
05		0.1		.05/.10	0.5	0.05	.1	0.3	0.1	0.3			<0.01	0.05	.001-0.5
0	1.0	5.0	2.0	5.0	15	0.1	5.0	1ppm	5.0	15.0	5.0		<1.0	5.0	<1.0-15.00
				<150	75	50	50		50	125			<50	150	<50-150
		500	240	250	300	250	250	200ppm	200	250	500		<250	500	150-600
7				1	2	.5	1					1	not detec-table		.5-17
		20	20	15	25	50	5	2	15	20	20	20	15	15	2-50
		1.5mg/L	3mg/L	<25	5	10	5	2	5	5	10	10	5	5	2-25
		2				5	5								2-5
			3°				5°	3°	3°				<75	200	75-200
													4	4	3-5
50	50	50	40	50	100	50	50	50	200	10	50	50	10	50	10-200
			6	10		.5	10		10		10	100	<10	<10	.5-100
10			50	200	20	50	50	0	10	10/20		100	10	200	10-200
100	100		40	50	50	50	100	100	100	500	50	100	<50	50	40-1000
50			4	2		.1		0			5	5			0-50
10	1		8	10		10	10		50	10	10	50	<10	10	1-50
			.25												.25-200
10			50	50	50	50	50	50	50	50	50	10/50	<50	50	10-50
		.2													.2
		500													500
		2000													2000
	4000			1000		100	1000		1000		1000		<1000	1000	100-1000

(e.g. International Drinking Water Standards). Such guidelines while relied upon in terms of their scientific validity, may need to be modified in the light of the local conditions and feasibility of implementation.

Similarly, national public or private bodies may recommend standards and institute 'seals of approval' for products which pass their tests.

Criteria Criteria reflect the state of scientific knowledge about environmental conditions or technical factors and their adverse effects on man and his environment. For example, air quality criteria might include the known properties of certain pollutants and the various ways of measuring them; a survey of present and past concentrations of the pollutants in the atmosphere; and a review of the evidence about the effects of various concentrations of these substances on man, animals, vegetation and materials, including epidemiological evidence. Such a criteria document sets out to be as objective as possible and does not in itself recommend, or set, standards.

In some countries, a deliberate separation is kept between criteria setting and standard setting, even to the point of assigning the tasks to different agencies. Criteria setting is seen to be a scientific, relatively value free process whereas standard setting is more constrained by the local political, economic and administrative setting as well as the values and perceptions of those who are defining the standard.

The degree to which standards are relative (despite their common appearance of being absolute) can be seen from the varying standards specified for the same pollutant or product by different countries or even different states or local bodies within the same country — sometimes even where the same criteria are used. For example, the scientific basis for drinking water standards is better established than for many other environmental hazards. Yet standards for different trace elements or for physical indicators such as mobility, taste and colour vary between countries by factors of 10 to almost 1,000 times (Table 5.5).

The tendency is for standard setting procedures to become better clarified and more open to wider scientific and public debate. Standard setting is a crucial aspect of risk management for it not only guides the subsequent regulatory and enforcement activities but also acts as an important monitor of political attitudes to the tolerance of risk. For example, some countries adopt much stricter standards for the control of highly toxic materials than others: the EEC Council of Ministers has agreed to the principle of 'no detectable emissions' of eight substances by 1980, but the UK has won a concession that resulting concentrations of these substances should be the guiding standard, not emission controls.

The actual procedures by which risk related standards are met are, therefore, a critical aspect of risk assessment. The principles involved here are:

(1) What is the role, composition and political effectiveness of scientific standard setting committees for various kinds of hazard in various countries?

(2) What role will standards play in relation to other risk management tasks? — a guiding role or an enforcing role?

(3) Will standards apply to ambient quality parameters as well as or instead of emission parameters?

(4) Will standards be subject to continuing review, both as to the effectiveness of their role and as to their scientific and political suitability?

(5) To what extent will independent scientific evidence be incorporated into the standard setting process, and that this be seen to be incorporated?

(6) Will the relationship between scientific criteria and subsequent politically established standards be made clear and subject to public discussion?

The answers to these questions vary from one kind of risk to another and from one country to another, but they should provide a guide for a national or international appraisal of the state of risk management.

5.3.4 Regulation and enforcement

Regulation is the process by which risk reducing standards or guidelines are applied in particular instances. In the case of toxic and biological risk, for example, this means the application of threshold limit values or maximum allowable doses which may be done on a provisional, presumptive or compulsory basis depending partly on the degree of cooperation versus adversary conflict between the regulator and the regulated that characterizes the risk management process.

Indeed, the regulatory role of standards and the comprehensiveness with which they apply are closely related to the degree to which the whole process is regarded as *cooperative arrangement* between risk producers, risk managers and risk receivers, or an *adversary relationship* among these three principal groups. In the cooperative approach, standards are established after a long period of consultation, objection and concession: the result is a voluntary or legally acceptable code of practice which is normally followed and usually is capable of being legally enforced.

The advantages are (a) friendly collaboration among all parties involved; (b) 'in house' confidentiality; (c) good working relations; (d) a shared commitment to steady improvement. The disadvantages lie in a certain exclusiveness in operation which may impede impartial scrutiny of all aspects of risk management, and which may foster a degree of decision control (sometimes referred to as non-decision making in the political literature) which could prove to have adverse consequences. It is virtually impossible to find out how far this kind of practice actually exists, partly because the whole process is so confidential, and partly because the participants themselves may not know how far they are controlling final decisions. But is seems that this kind of activity could be most prominent in precisely those areas where scientific and political controversy is greatest.

These arguments fall fair and square on the 'best practicable means' approach adopted in the UK and many other English speaking countries: this is widely regarded to work well most of the time, but may not always prove to be the most suitable managerial principle in all cases of risk management. The adversary procedure is more commonly found in federal states where the law is constitutionally strong and where distrust of regulatory procedures has a long political history. Its chief advantages lie in its precision and ease of enforcement, while its disadvantages relate to its unworkability in the face of conflicting political demands and to its general rigidness and inflexibility. In practice, an amalgam of the two approaches, leaning to one side or the other depending on circumstances is followed in many countries, the exact regulatory enforcement varying by class of risk and political 'style'.

The major questions relating to regulatory procedures apply to the sincerity with which the whole process is conducted and the degree of independent analysis available. These are not readily testable, but they are very important. To what extent are regulators and regulated genuinely committed to achieving socially accepted levels of protection against risk? In short, does their motivation come from within or without? This self-regarding social ethic toward risk protection is relevant through the whole of the risk management process, especially in regulation and enforcement. To test for this is bound to be a difficult task but not an impossible one, though it should be carefully controlled.

Why should the decision-maker be concerned with this? Because successful risk management can depend critically on its accountability and ability to hold up under scrutiny, and the more society knows it can trust its officials the better they will be able to continue their roles in the future. Some criteria for assessing accountability are:

(1) The degree of access to risk management officials by independent experts and responsible media. This may be by formal and informal means.

(2) The character of devices employed to ensure that all relevant viewpoints are heard and shown to be taken into account. These should ideally be based on some kind of dialogue to permit antagonists to challenge each other freely.

(3) The nature of information dissemination, especially as to why and how standards are established. This can be based on the publicity as to reasons for reaching decisions plus the amount of public answerability to independent, but scientifically respectable monitoring groups.

The enforcement of regulations by individual government inspectors or agencies also depends very much on the managerial 'style' of the individual or agency, and these can show wide variation even within one country. Some other managerial characteristics also influence how effectively regulations are enforced. These include:

(1) The degree to which political judgments can knowingly permit an overriding of regulatory standards in particular cases.

(2) The constitutional role of the law and the courts in enforcing good environmental practice and specific legislation.

(3) The legislative intent towards tough enforcement operations, defined by such criteria as the level of fine permitted and actually imposed, and the status of the reviewing body (magistrates court, county court etc.) for various categories of offences.

(4) The relationship between public scrutiny and official monitoring will influence enforcement in certain risk areas (e.g. nuclear-related risks) if public risk and anxiety is especially high.

5.3.5 Emergency response

Not all events lend themselves to an orderly decision control process. Events occur, which because of the severity of impact require immediate response — in other words, create an emergency. In emergency situations the normal decision-making process must be suspended and emergency authorities invoked.

In the United States, for example, the legislation for the various environmental programs give the Administrator of the Environmental Protection Agency the authority to declare a particular event or espisode an emergency, which permits him to invoke the emergency permitted under the Act.

Such emergency procedures permit the Agency to immediately

(1) Suspend activities.

(2) Ban products and withdraw them from circulation.

(3) Establish standards, criteria or regulations based on existing knowledge and *without benefit of the normal review, comment and concurrence procedures.*

(4) Authorize remedial or preventive measures.

Two examples of fairly recent emergency situations in the United States are (1) the air pollution episode in Pittsburgh, Pennsylvania in November 1975, and (2) the Kepone incident in Hopewell, Virginia in July of that year.

In the Pittsburgh case, EPA officials were advised by the Municipal authorities that air pollution levels were getting dangerously high. The Administrator ordered an EPA medical doctor to the scene and he confirmed levels were dangerous to health. On the basis of his recommendation the city ordered major polluting industries to close down, suspended school and invoked a no-driving ban for the duration of the emergency.

In the Hopewell Kepone incident, tests run by EPA and other Federal and State agencies confirmed that Kepone levels in the James River posed a threat to human health leading to closure of the river to commercial fishing by the Governor of Virginia and the setting of emergency permissible levels of Kepone in shell and fin fish by EPA.

It is important for risk managers to be able to act quickly when emergencies arise and special provisions need to be considered in drafting legislation, in delegating agency functions and in designing risk management procedures.

CHAPTER 6

International Collaboration in Risk Management

In the assessment and management of environmental risks, 'no man is an island' and nor is any nation. All assessments are made in the context of a wider community that creates greater opportunities than the individual person or single nation would have acting alone. The wider community also imposes constraints and creates areas of potential conflict.

The purpose of this concluding chapter is to show, partly by example, how risk assessments and their implementation enter into the relations between nations, and to point out needs and opportunities to harmonize the actions of individual nations and make them more mutually supporting. Indeed a reason for developing the tools of risk assessment and moving towards a common understanding of their use is precisely to make international collaboration more feasible.

There are three main reasons which make international collaboration useful and indeed essential in the management of environmental risks. First, some of the problems are transported across international boundaries by environmental processes and affect contiguous nations, or groups of countries, and in some cases the whole world. Second, some environmental management decisions taken in one country have repercussions in others because they are economically linked through trade or international aid programmes or simply because of the dissemination between scientists of different countries of information about scientific observations on risk. Thirdly, just because the community of nations is so strongly knit together by environmental and economic links the major problems of one nation are inevitably in some degree the concern of all. Food shortages caused by drought in one country cannot and should not be ignored by others simply because they are not directly affected.

6.1 ENVIRONMENTAL LINKS

The environmental processes which link nations occur on very different scales of space and time. Some are quite limited in extent and concern only two contiguous states. This is often the case, for example, when two countries share a river basin as the United States and Canada share the Great Lakes — St. Lawrence Basin. Other environmental processes are regional in scope. The

Sahelian drought affected a group of countries extending from the Red Sea to the Atlantic, all the way across the southern margins of the Sahara Desert. Yet other environmental processes are global in extent and implications. DDT residues have been found in Antarctic penguins, and ozone-layer depletion potentially affects the earth's atmosphere above every country.

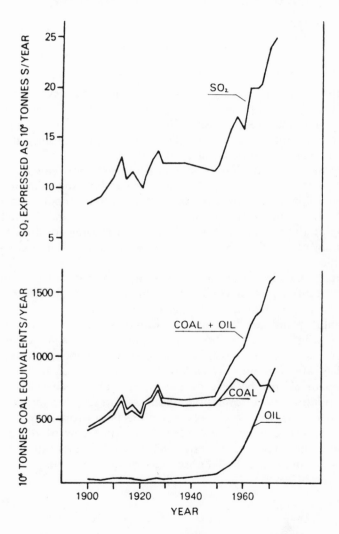

Source: Fjeld, 1976. (Reproduced by permission of the Norwegian Institute for Air Research)

Figure 6.1 Fossil fuel consumption and estimated anthrogenic SO_2 emissions in Europe, 1900-1972.

6.1.1 Trans-border Problems

Acid Precipitation For some time the major industrial nations have sought to reduce the health risks associated with sulphur dioxide and particulates in the atmosphere by building taller chimneys and smoke stacks at industrial plants. This has served to disperse the effluents over wider areas and thus reduce concentrations near the ground in populated and heavily industrialized areas.

The risk management policy adopted by individual nations has been successful in that in many localities *peak* concentrations of sulphur dioxide have been reduced. As shown in Figure 6.1 the total volume of sulphur dioxide emissions in Europe have continued to increase rapidly. In recent years it has come to be recognized that long-range transport of sulphur dioxide can have detrimental effects at great distances from the source. Sulphur dioxide is readily converted to sulphate (SO_4) aerosol form and these particles are carried downwind from industrial areas. There has been widespread acidification (lowered pH) of rain and snow in Europe and eastern North America from this source (Ambio, 1976). Truly global dispersion, however, is not achieved; some Southern Hemisphere soils even show sulphur deficiencies. There are strong indications that the increased acidity of precipitation is a main cause of the extensive fish kills observed in southern Scandinavia.

A similar phenomenon is occurring in the northeastern United States and adjacent parts of southern Canada. The problem is exacerbated where pollutants may accumulate in the snow cover during the winter season. This may give rise to high concentrations in the first meltwater and to sudden increases of acidity in exposed areas.

In large areas of eastern North America and Western Europe the risk now exists that many thousands of lakes will become biologically unproductive — in other words dead — if emissions of sulphur dioxide continue.

A study conducted for OECD on the Long Range Transport of Air Pollutants (LRTAP) has produced data showing the Europe-wide concentrations for sulphur dioxide (Figure 6.2) and has permitted the compilation of a table showing sulphur emitters and sulphur receivers (Table 6.1).

Clearly sulphur dioxide emissions have become in Europe a problem of trans-border dispersal, and international cooperation will be needed. Relocation of industry on the scale required is hardly a practicable solution. The reduction of emissions under certain weather conditions might be possible but it would require agreement on selection of the areas to be protected. The obvious control strategy for the regional air pollution problem seems to be a reduction of emissions primarily in large cities and in heavily industrialized regions. Statements of principle have been adopted at OECD and other international organizations concerning the necessity to reduce emissions.

How this would be carried out technically and at what cost has not yet been determined. Those countries contributing most on the emissions side are worried about the expense involved and tend to suggest alternative approaches

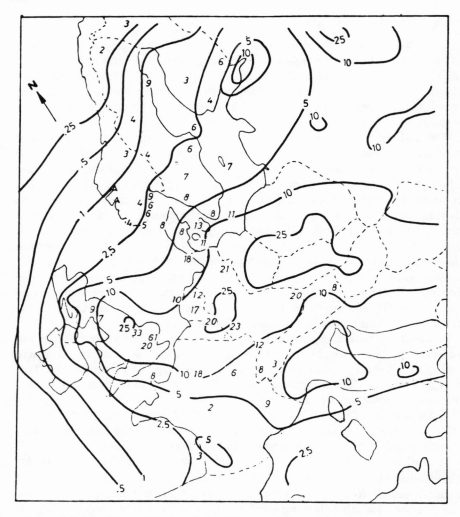

Source: OECD, 1977, p.9-7. (Reproduced by permission of OECD).

Figure 6.2 Estimated mean concentration field for SO_2 for 1974. Observed mean concentrations given by italic numbers. Unit $\mu g\ SO_2/m^3$

such as liming of rivers and lakes in Scandinavia. The Scandinavian countries point out that they have already reduced their own emissions considerably and that this is possible for other countries. Liming of the large areas affected involves considerable practical problems. The Scandinavian countries are now calling therefore for an international agreement to reduce air pollutants in Europe (Ottar, 1977).

It is a characteristic of trans-border dispersal problems that the nations where the origin of the problem lies are reluctant to spend money solely to

Table 6.1 Estimated Budget for Total Deposition of Sulphur for 1974

Rows = Emitters; Columns = Receivers. Unit: 10^3 tonnes S.

Emitters \ Receivers	Austria	Belgium	Denmark	Federal Republic of Germany	Finland	France	The Netherlands	Norway	Sweden	Switzerland	United Kingdom and Ireland	Czechoslovakia	German Democratic Republic	Italy	Poland	Other areas	Undecided	Sum	Annual emission
Austria	60	6	0	40	0	20	2	0	0	5	20	20	20	30	7	20	30	300	221
Belgium	0	100	0	20	0	30	5	0	0	0	30	1	4	0	0	1	10	200	499
Denmark	0	1	60	6	0	3	1	0	2	0	10	1	6	0	2	2	10	100	312
Federal Republic of Germany	8	60	7	700	0	100	40	0	2	7	100	20	80	7	10	10	90	1300	1964
Finland	0	2	8	10	100	4	2	2	30	0	10	7	30	0	20	80	70	400	274
France	2	40	1	50	0	600	10	0	0	6	100	5	20	30	2	30	150	1000	1616
The Netherlands	0	10	1	10	0	10	60	0	0	0	30	1	4	0	1	0	10	150	391
Norway	0	4	8	10	1	9	4	30	9	0	60	3	10	0	5	4	100	250	91
Sweden	0	7	30	30	10	10	6	6	100	0	40	8	50	6	20	30	100	500	415
Switzerland	1	2	0	7	0	20	1	0	0	30	10	2	1	0	1	2	20	100	76
United Kingdom and Ireland	0	8	2	10	0	20	4	0	0	0	800	2	9	0	2	1	100	1000	2883*
Czechoslovakia, German Democratic Republic, Italy, Poland and other areas	60	60	80	400	40	200	40	9	50	10	600	900	1000	900	4500	1000	1000	11000	—
Sum	100	300	200	1300	150	1000	200	40	200	60	1800	1000	1500	1100	4600	190	1900	17000	—

*including 80 × 10^3 tonnes S from Ireland

Numbers are rounded to one significant figure and accurate to within ±50 per cent. The sums are calculated from unrounded figures and thereafter rounded separately.
Unit: 10^3 tonnes S.

Source: OECD, 1977. (Reproduced by permission of OECD)

protect their neighbours, and that for the most part the recipients of the problem can do little about it.

Many other examples of trans-border problems could be cited. These include, for example: the high levels of sulphur dioxide and particulates in Windsor, Ontario, Canada originating in the industrial areas of Detroit, Michigan, USA across the river; the pollution of the lower Rhine in Holland with effluents from German and Swiss industries upstream; the development of locust swarms or other pest infestations in one country and their migration to devour the crops in another; the construction of the Farrakka Barrage on the lower Ganges by India diverting water into the River Hooghly which would otherwise flow on into Bangladesh.

These are all examples of a natural environmental process which transfers a risk from one country to another. They are confined to problems involving adjacent countries, or those close linked in space. Their solution commonly requires bilateral agreements as in the work of the International Joint Commission involving the United States and Canada in the management of the Great Lakes, or regional groupings such as the member states of the Tchad Basic Commission (Cameroun, Chad, Niger and Nigeria). Contiguity or close proximity on the same continental land-mass is a necessary condition for these effects. Bilateral and multilateral agreements can be reached and agreement is generally facilitated if scientific studies of the risks involved have been carried out on a collaborative basis. The negotiations can proceed more effectively if there is agreement on the scientific character of the problem. The OECD study of acid rain in Europe is an example of scientific cooperation preceding the negotiations on the management decisions required.

Desertification The recent Sahelian drought in Africa from 1968-75 has served to focus worldwide attention on the problems of desertification. Climatic fluctuations exacerbated by resource exploitation activities such as deforestation, overgrazing, deep-well drilling and firewood gathering have now affected at least half of the countries in the world and 30 per cent of the world's land surface with desertification. Many of those affected are amongst the poorest nations and lack the economic resources to combat desert encroachment.

About sixty million people live on the margins of present deserts and are subject directly to risks of drought and desertification. It has been suggested that 6.7 per cent of the world's surface is *man-made* desert caused by deforestation, overgrazing, burning and farming. This is an area larger than Brazil (Kassas, 1975). In the southern Sahara alone 650,000 square kms of land suitable for agriculture or intensive grazing have become desert over the past fifty years (US AID, 1972). Elsewhere deserts are also expanding. In northern Chile, the Atacama Desert is advancing at 1.5 to 3 kilometres per year and the Thar Desert in India was estimated during the 1950's to have been enlarging at the rate of 0.8 kilometres a year for fifty years (Roy and Pandey, 1970).

Efforts to improve man's livelihood in wet years can increase vulnerability to drought and thus speed desertification. Almost all these efforts face the risk that by increasing population density and wealth, desertification will intensify.

The international collaboration required in the field of desertification includes financial and technical assistance to the countries most severely affected, and the exchange of knowledge and experience in methods of combatting desert encroachment. Collaborative research is also needed on an international basis.

6.1.2 Global Hazards

Acid precipitation and desert encroachment are largely regional in their effects. Other risks are globally distributed. The atmosphere's remarkable assimilative capacity distributes certain pollutants around each hemisphere quickly (because the winds are mainly from west or east), and from pole to equator within six to twelve months. Some pollutants — for example the oxides of nitrogen, NO_x — are quickly removed by falling rain. Others break up chemically into harmless materials. The larger particles fall out gravitationally, are scavenged by precipitation, or dissociate chemically. But there remain certain pollutants that resist all these cleansing mechanisms, and are hence diffused globally.

Among these, for example, are the stable gas sulphur hexafluoride, SF_6, which is released from modern electrical switching gear. It has been detected in small concentrations from pole to pole, and high into the stratosphere. It has no known sink. There are many similar stable gases in the atmosphere undergoing this type of global dispersion. They do little harm if (i) they are optically neutral, and hence do not disturb the earth's radiation balance; (ii) they are non-toxic to man and biota; and (iii) they are chemically and photochemically stable and inert (i.e., they resist dissociation and reaction with other species). Sulphur hexafluoride meets all of these criteria. Unfortunately many other pollutants do not.

Carbon dioxide, for example, is being added to the atmosphere at an alarming rate. About four billion tonnes come from the burning of fossil fuel, and an amount variously estimated from two to eight billion tonnes is added through the destruction of forests, and the oxidation of soil humus. The oceans may absorb three billion tonnes by solution. The rest stays in the atmosphere, whose carbon content (near 700 billion tonnes) is rising at the rate of 3 per cent per decade. If the trend continues, atmospheric carbon content will double by about 2050 AD. This should have a significant warming effect and could cause major readjustments of world rainfall distribution (Perry and Landsberg, 1977; Keeling and Bacastow, 1977; US National Academy of Sciences 1977a; Woodwell, 1977; Kellogg, 1977), with unascertained social and economic impacts. To continue burning fossil fuels at present or increased rates thus poses a serious risk to future climatic stability and to climate-sensitive activities.

A similar dispersion has been achieved for the nine million tonnes of chloro-fluoromethanes (CFMs) so far manufactured and used as spray-can propellants, plastic inflaters (polyurethane foam) and refrigerants. These gases are as widely dispersed as SF_6, but are dissociated above 20 km by hard ultraviolet radiation. The freed chlorine then attacks ozone (O_3), allowing a greater penetration to groundlevel of harmful ultraviolet-B radiation (wavelength 295-320 nanometres). It is possible that nitrous oxide from heavily fertilized land adds to this effect. A reduction of ozone of order five to ten per cent has been predicted in computer modelling exercises, and this may lead to a ten to twenty per cent increase in the incidence of skin cancer among fair-skinned people (USNAS, 1977b; Evans and Hare, forthcoming, 1979). However, the natural variability in stratospheric ozone amounts is so large that there is as yet no evidence for a downward trend, even though concentrations of chlorofluoromethanes have been increasing.

It is not yet possible to construct a table for the ozone depletion problem similar to the table of emitters and receivers of sulphur dioxide in Europe (Table 6.1). Development of such a balance sheet is fundamental to the attainment of any international agreements which may become necessary to safeguard the ozone layer. An early impression was created that the ozone problem was and would remain the concern of the larger industrial nations. A substantial proportion of the world's production of fluorocarbons is concentrated in the United States (Machta, 1976; Munn, 1977). Currently the sale of spray cans using fluorocarbons as propellants is banned in the United States, but not their manufacture or export. Total world use has continued to increase despite the ban in the United States.

Recently the possibility that nitrogen fixation may contribute to ozone depletion has been recognized. Natural processes fix nitrogen in the soil and air but man's contribution to the total is increasing (24 per cent of the total in 1974). In 1850 no industrially produced fertilizers were used; in 1950 3.8 million tonnes, or two per cent of the total, came from industrial fixation. If the growth in nitrogen fertilizers continues, the amount of nitrogen fixation by fertilizers alone may increase to between 100 and 200 MT/yr by the year 2000. It is denitrification which produces nitrogen oxide, N_2O, a gas capable of reaching the stratosphere. During denitrification of the soil, only a small and uncertain fraction (about 7 per cent) is believed to be converted to N_2O. The processes of denitrification in the soil and the sea are not well understood. In particular there is a lag between application of nitrogen fertilizer and denitrification which has been estimated as possibly lasting hundreds to thousands of years (Lin *et al.*, 1976).

It is possible that nitrogen fertilizers will within decades become the major single source of atmosphere N_2O. Since all countries are potential users of nitrogen fertilizers and since rapid increase in their use is anticipated, especially in developing countries, it is clear that atmospheric ozone depletion is potentially at least a a matter of concern not only to industrial nations but to all countries of the world, especially perhaps the larger agricultural producers.

Research on the ozone layer continues to expand under international auspices (UNEP, 1978) and a World Plan of Action has been formulated (UNEP, 1977).

Global hazards differ from trans-border dispersal problems in degree only. They concern those natural environmental processes which can become so pervasive that the common property resources of the entire global community are affected. The main common property resources are the atmosphere, the oceans and outer space. Man-induced changes now threaten the stability of the world's climatic zones (Hare, 1977).

Marine oil spills Contamination of the marine environment by petroleum hydrocarbons is one of the major ecological problems facing the world community today. A detailed report by the US National Academy of Sciences (1975) estimated that 6.1 million metric tons of oil enter the oceans annually. The distribution of massive oil spills 1967-77 resulting from accidents involving tankers or ocean oil wells, is shown in Figure 6.3.

The causes of this problem are two-fold — the heavy reliance on enormous quantities of imported petroleum by the western industrial nations and the increasing size of oil tankers. For example, US imports increased from 3.4 million barrels per day in 1970 to 8.3 million barrels a day by 1977. Both the size and number of tankers has increased. Before 1950 there was only one tanker above 50,000 dwt. By 1965 there were 47 tankers in the 50,000 dwt. to 200,000 dwt. classes, and by the late 1960s there were 131 tankers above

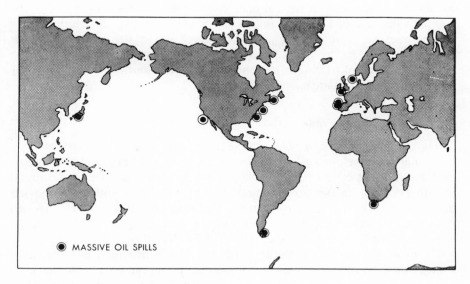

Source: Grundlach, 1977.

Figure 6.3 Distribution of massive oil spills in the world 1967-78

200,000 dwt. in operation. Both the fleet and the size of individual tankers continues to grow. Massive spills have therefore become inevitable and seem bound to increase.

From all the major incidents, experience has shown that impact is highly variable. Coastal zones are likely to suffer severe damage to wildlife, shellfish production and to recreational amenities. Spills offshore cause only minor environmental impairment (Grundlach, 1977).

Fish depletion Between 1950 and 1970 the world fish catch rose from 21 million to 70 million metric tons. It grew at over five per cent annually which was twice as fast as the world population and gave hopes for a solution to world protein shortages. By the 1970s however serious stresses in the marine resources were seen and for the first time, the total global fish catch maintained a decline over several years. Overexploitation had wiped out some species of fish and depleted many others to a fraction of their potential yields (Eckholm, 1976, pp.155-6).

6.2 SOCIO-ECONOMIC LINKS

The socio-economic connections between nations grow steadily stronger and more pervasive. It is a well established axiom of international trade and finance that economic weakness or collapse in one nation works to the detriment of all. Such a strong recognition of interdependence has not yet been achieved in environmental affairs. National self-interest still outstrips collaborative action in the case of many environmental risks. The risks persist not only because of scientific uncertainties and technical difficulties. They persist because nations have failed to agree or to perceive a need for agreement. It is in the realm where environmental risks intersect with the socio-economic links between nations that the contributions of scientifically based risk-assessment are more sorely needed.

6.2.1 International Trade

Pakistan exports mango pickles to the United Kingdom. This is a flourishing trade that has developed over the past 15 years. In 1973 the United Kingdom health authorities conducted a test on a shipment of mango pickles and found quantities of lead in excess of the maximum permissible amount of two parts per million. The whole consignment was refused entry into the United Kingdom and was returned at the expense of the exporter to Pakistan where it was subsequently marketed and consumed (SCOPE, 1977).

The exporter approached the Pakistan Council of Scientific and Industrial Research Laboratories to find which ingredient in the cargo had contained too much lead. Each ingredient was analysed and the lead was eventually traced to turmeric. Turmeric from a number of sources and regions in Pakistan was then analysed and it was found that lead was present in large quantities in those samples of turmeric grown near cities and not far from roads. Turmeric from the more inaccessible rural areas was uncontaminated. The PCSIR considers that lead from petrol used by automobiles in the more urbanized regions of the country is the source of lead in tumeric and in mango pickles. Care is now taken to use turmeric from the rural areas in the manufacture of mango pickles. The exporter has also recognized the need to meet standards set in countries to which his products are exported and now has a small quality control laboratory for testing purposes (SCOPE, 1977).

The standard for lead in food in Pakistan is the same as that in the United Kingdom. The adoption of a standard does not necessarily mean, in Pakistan as elsewhere, that it will be enforced. The monitoring and inspection required to ensure enforcement of standards is expensive (Beg, 1977).

This example is a small illustration of a widespread and general problem in risk management. Nations for whom international trade is an important factor in their economic health and development are naturally concerned about the effects that environmental risk assessment decisions may have on their international trade position, on their national income and on the prosperity or viability of specific economic activities, such as manufacturing, resource development, urbanization, and public services such as water supply, transport, and so on.

Fears are most frequently and strongly expressed about the effect of environmental standards which if imposed in one country and not others may result in adverse consequences for an industry, for the nation's balance of payments, for real income levels and for the nation's long-term comparative advantage. This sort of fear is especially great where other (competitor) countries adopt lower environmental standards, or directly or indirectly subsidize the cost of environmental quality controls.

Clearly a degree of international collaboration is required if major trading nations are to continue to take steps to reduce the risks of environmental contamination resulting for example from industrial pollution. A difficulty in this process is that the adoption of the same standards or the same degree of risk can have disproportionate effects in different economic circumstances. The requirement to meet a particular emission standard for automobiles is more difficult to achieve in some makes of cars than in others, for example. The problem of producing cars (or other goods meeting different standards for different markets) complicates production runs and disadvantages some more than others.

Such effects clearly create the possibility that decisions taken in the interests of risk reduction or safety may in fact be used as restrictive trade practices. It is reported for example (d'Arge and Kneese, 1972) that the French government requires inspection of production procedures for quality control of pharmaceuticals and that this has effectively frozen foreign manufacturers out of the French market.

Another example is a law in the Federal Republic of Germany adopted in 1974 restricting the lead content of gasoline to 0.15 grams per litre (Sterling,

1971, quoted in d'Arge and Kneese, 1972). Ostensibly the law was passed to reduce lead emissions from automobiles in West Germany, but it happened to have the effect simultaneously of restricting sales of other countries' automobiles in West Germany — clearly a non-tariff trade barrier. West German vehicles (in 1971) could be easily adjusted to low lead gasolines but several major Italian and French high compression engines could not be so easily adjusted.

It is not known whether this decision was in fact aimed at creating a non-tariff barrier — but it indicates the difficulty in knowing which decisions ostensibly made to reduce risk are really made for other ends.

Trade regulations can also have the effect of 'exporting' risks to other countries. Under proposed changes in the trade regulations for drugs in the United States, overseas exports of US-manufactured drugs would be permitted even though the drugs have not been approved for sale in the US. This proposed policy change rests on the fact that different countries have different public health problems, and that a drug with an unfavourable benefit-to-risk ratio in the US might have a favourable ratio in another country. Such a change puts the onus on the receiving country to assess the risks and benefits of drugs in terms of its own circumstances.

Nations share a common interest in collaborating to reduce environmental risk. Where risk management conflicts or appears to conflict with national economic interests in international trade scientifically based risk assessments have a key role to play in laying a basis for common understanding.

6.2.2 International Aid

Dutch engineering skills, supported by financial resources raised in Washington, D.C. assisted in the construction of dykes and cross-dams in the coastal delta areas of Bangladesh. These projects carried out in the 1960s provided some flood protection for the population living on the deltaic islands or 'chars'. The cross-dams also accelerated the siltation process thus making more land available for settlement and cultivation. The development aid thus provided helped Bangladesh (then East Pakistan) to produce more food and find gainful agricultural employment for more people.

Disaster stalked behind these well designed and well intended efforts. The tropical cyclone of November 1970 drove water inland almost seven meters above normal high tide. The tidal onslaught struck the outer chars about 11.00 p.m. on 12 November 1970 and by daylight at least 225,000 people were dead, ripening crops worth $63 million were destroyed and 280,000 head of cattle were swept away (Frank and Hussain, 1971, p.439).

This high death and damage total occurred in spite of the fact that the storm had been identified three days earlier and tracked by satellite on its course northwards up the Bay of Bengal. Warnings reached the capital at Dacca but were not effectively communicated to the rural populations living in the most exposed areas. Even if warnings had reached the victims, with the limited transport available, few could have escaped.

Seeking to improve the health and productivity of the cattle-rearing pastoralists of sub-Saharan African, French and other countries' technical assistance carried out a programme of drilling deep tube wells, and reducing animal diseases during the 1960s. Cattle herds increased in size and some health improvements were achieved. When the Sahelian drought of 1968-75 struck the region there was insufficient production from the grasslands to support the increased number of cattle. The land was badly overgrazed, especially in the vicinity of water sources, thousands of people died of starvation or malnutrition-related diseases, and many cattle were lost. People were forced to migrate south in large numbers. An estimated two million people fled to refugee camps (Newman, 1975). The overgrazing led to an expansion southwards of the desert zone. The process of desert advance will be difficult to halt and almost impossible to reverse in the short-run.

Other factors involved in the process of desert advance were that governments and international aid agencies showed a lack of understanding of the traditional ways in which the pastoral nomads were in equilibrium with their environment and a lack of perception of the environmental risks to which the populations were subjected (UNESCO, 1975).

These are two of the more dramatic examples of disaster following in the wake of development assistance. They serve to emphasize that technology and management methods should not be blindly transferred to other countries without a good understanding of the differences in the natural environment and in the society. Risk assessments need to be made in light of the circumstances of the recipient country and not simply transplanted from the donor country.

Bangladesh cyclone and Sahelian drought disasters are examples of natural hazard and resource depletion risks exacerbated by external aid programmes. Disasters on a similar scale have not yet occurred in pollution risks or technological risks. As the scale of industrial development expands however, it appears likely that similar catastrophes will occur in other risk areas. An indication of the sort of thing that can happen is seen in the Iraq mercury poisoning epidemic of November 1971-January 1972. Imported grain that had been intended for use as seed grain was made into flour and used to supply bakeries. Unfortunately the seed grain had been treated with a mercury fungicide. The first cases of mercury poisoning were admitted to hospital in November 1971, and by January 1972 hundreds of cases were being identified daily. Once the problem had been tracked to its source the epidemic was brought to a swift end and no more cases were reported after March 27, 1972.

This was the most catastrophic epidemic of chemical poisoning so far recorded. A total of 6,530 cases were admitted to hospital, and 459 hospital deaths were attributed to methyl mercury poisoning. It now seems probable that at least 3,700 people died in the epidemic and that more than 10,000 became ill, many of whom may be permanently impaired (Ontario, 1976).

Industrial expansion in developing countries and the rapid growth in the use of fertilizers, pesticides and other chemical products in agriculture often stimulated by external assistance, will mean the creation of a new set of risks.

138

The monitoring, testing, screening processes which accompany these risks in the industrial nations are expensive and require much scientifically trained manpower. Since both money and trained manpower are in short supply in most developing countries and there is a temptation to seek the short-run benefits and to discount the new risks so generated.

Where development assistance is involved in projects that create new risks it is important that the recipient country be fully informed of the character and possible consequences of the risk. It is also important that neither country should automatically assume that the same standards of safety will apply.

The risks attendant upon development activities must be more fully examined and understood by both donors and recipients. The decision on the acceptability of the risks or the safety levels or standards to be attained is primarily that of the recipient country.

The risks are not always obvious. In the case of the Bangladesh cyclone disaster, the Sahelian drought and the Iraq mercury poisoning, the full risk system in all its ramifications was not explored. To the extent that the risks were known and understood by experts, their knowledge was not effectively communicated or acted upon.

6.2.3   Risks to Internationally Valued or Unique Sites

The Government of India is building a large oil refinery in Math'usa some 30 miles from the Taj Mahal. Sulphur dioxide emissions from the refinery combined with water vapours in the atmosphere will form an acid rain, that if present in sufficient quantity over a period of time could react with the marble (calcium carbonate) structure of the Taj. The white polished surface of the Taj Mahal, famous for its luminescence in the moonlight, could become discoloured, then pitted and scarred.

Told in this fashion the story suggests a cause for international concern. The Taj Mahal belongs to India, but also in a wider sense it is part of the common cultural heritage of mankind. If the Taj were to be irreparably damaged, the world and not just India would be the poorer.

A risk assessment has been carried out. It involves forecasts of the amount of sulphur dioxide to be emitted by the refinery under several alternative control technologies. It involves atmospheric diffusion models to calculate how much of the emission would reach the Taj. There is a low inversion layer in the area most of the time for a 6-8 months period. The prevailing winds are northeasterly for nine months of the year, which places the Taj exactly downwind of the refinery.

There is little reliable information on the chemical reaction rates between the marble used in the Taj and SO_x. Inferences can be made from some Swedish data about limestone and SO_2.

If control equipment is used, and if low sulphur-content crude oil is used then the additional SO_2 at the site of the Taj Mahal is calculated to be in the order of $0.1 \mu g/m^3$ (micrograms per cubic metre). From estimates of reaction rates it appears that no damage would occur at levels as high as $20\mu g/m^3$. From observation in the vicinity of the Taj Mahal however, existing levels are already in the vicinity of 40μ g/m^3 due to numerous small scale industries and other sources in the vicinity of Agra.

The conclusion of the risk assessment therefore, is that there is a risk to the Taj Mahal, but that the contribution of the proposed refinery is very small in relation to the existing levels of pollution from other sources. The management implication drawn by the study and accepted by the Government of India is that the refinery should be built on the approved site, and that emission control technology and low-sulphur content crude should be used. At the same time it is evident that a substantial clean-up effort is required for the other, already existing sources of SO_2 in the region, if the Taj Mahal is not to suffer damage.

If such action is taken it will reduce the risk to the Taj Mahal, but not eliminate it. Is the level of risk acceptable? The Archaeological Survey of India and some Indian scientists are reported to be concerned (Sri Vatsa, 1977). They are not convinced that the safeguards are adequate or will be adequately enforced. Indian industrial licensing laws, they point out, do not include pollution control standards. The treasures of the Taj cannot be moved; the site of the refinery can be. Not only is the debate being conducted in India, but the opponents of the refinery have formed an International Action Committee which has appealed to scientists and intellectuals throughout the world for help in persuading the Government of India to relocate the refinery.

The appeal is based on the notion that the Taj Mahal is more than an Indian architectural site. It is argued that all people have an interest in the preservation of the Taj. Such a view led UNESCO to lead an international effort to save Egyptian archaeological sites from submergence under the waters of the High Aswan Dam. Similarly concern has been expressed in many countries for the preservation of Venice (slowly sinking) and the Acropolis in Athens (being corroded by air pollution).

When development involves the creation of an environmental risk to an internationally valued or unique site, nations may expect other countries to take a close interest in the safeguarding of such monuments or sites. There are already examples where the principle of international cooperation has been accepted. So far this has always been with the approval of the country in which the monument or site is located. If, in the interests of development any country wishes to place at risk part of the common cultural heritage of man within its own borders, other nations and peoples have no recourse but to moral persuasion.

6.2.4 Exchange of Information and Scientific Knowledge

When the United States decided on 18 October 1969 to impose a complete ban on all uses of cyclamates (artificial sweeteners) the action was swiftly repeated in more than 30 other countries. For example, Sweden took action on

20 October. Decision in developing countries generally came a little later. For example, Uruguay took action on 4 November, Nicaragua on 20 November, Ethiopia on 2 December, Saudi Arabia on 26 January 1970, Philippines on 4 August 1970.

When Canada and the United States announced a similar decision about saccharin in 1977 it was not widely followed and seemed to have been largely rejected outside North America.

The first cases of Minimata disease, an epidemic neurological disease, occurred in Japan in 1953. It was not until 1959 that methyl mercury was positively identifiied as the cause. By 1964 mercury seed dressing had been implicated in the poisoning of seed-eating birds in Sweden (Löfroth, 1970). In 1962 scientific papers on the Japanese experience were published in English for the first time identifying mercury-containing industrial waste-water as the cause of Minimata disease. Five years passed before monitoring of mercury levels began in Canada in 1967 and the first research results did not appear until 1969 (Fimreite, 1969). High levels of mercury were found in Ontario fish leading to the suspension of commercial fishing in some waters.

In Nigeria there are no regulations or standards concerning the import, sale or use of chemical pesticides. Manufacturers can sell pesticides in Nigeria even if they have been banned or had their use restricted in the country of origin. The level of pesticide use in Nigeria is relatively small today but it can be expected to increase rapidly in future.

The dissemination, use and exchange of scientific information about environmental risks is an extremely haphazard process. In some cases information that could be beneficially shared appears not to reach those who could use it, or to do so only after long delays. In other instances new information from laboratory tests or experiments is followed by swift action in a number of widely separated countries. On other occasions similar information goes largely unnoticed or ignored.

Sometimes it appears that there is too much information for some governments to absorb or react to. For example, the information generated by the US Environmental Protection Agency is too voluminous for most countries to be able to decide what applies to their own situation. This problem of information overload is likely to be further exacerbated by the large volume of scientific data that will be generated under the US Toxic Substances Control Act of 1976. Many countries will lack the scientific manpower or the funds to make decisions about what, if any, of it is relevant to their own circumstances.

Much of the scientific information about pollution risks that has emerged in the past has not been important to many developing countries. The scale of industrial development proceeds however, and as the use of chemicals expands in agriculture, there will be increasing need for developing countries to be aware of the new risks that are being introduced. The training of manpower in the environmental sciences should accompany and not follow the environmental inputs of development; and more consideration needs to be given to the international dissemination of environmental risk data and the capacity of countries to judge it in the light of their own circumstances, and to use it effectively.

6.3   COMMON NATIONAL PROBLEMS

A third reason for international collaboration in risk management is simply that many countries face large problems which are essentially similar to those found elsewhere. The environmental risks generated do not necessarily cross international boundaries, nor are they necessarily involved in the socio-economic links between countries. There is clearly a class of environmental risks however, about which it is helpful to share information, scientific knowledge and management experience.

An example of a common national problem of this sort is water supply. According to World Health Organization statistics there were 1,026 million people living in rural areas in 90 developing countries in 1970 without access to safe supplies of drinking water (Burton, 1977).

The risk to health from bacteriological contamination of drinking water has been recognized as a priority problem and the period 1980-89 has been designated the International Drinking Water Decade. A variety of activities is planned for the Decade including a major expansion of development assistance.

There are clear advantages in the international recognition of such problems and the development of concerted action, even though the problems themselves are unconnected by environmental or economic links.

Other similar problems include malnutrition, poor housing and sanitary conditions, excessive noise, soil loss and land transformation, deforestation and natural hazards such as drought, earthquake, flood or tropical cyclones.

The establishment of procedures for collaboration in these areas may set helpful precedents for later joint action to deal with industrial pollution and global risks of climatic change and the like.

6.4   EMERGING NEEDS AND SUGGESTED ACTIONS

Not much more than a decade ago the work that is now being done by international environmental organizations like SCOPE and UNEP would have been considered utopian. Indeed the very existence of these and other such organizations was but a dream. That they now exist, are well established and are doing work to some good effect, is testimony both to the continued adaptive capacity of human institutions and to the seriousness of the environmental risks which rush upon us.

The adequacy of existing institutions to deal with a growing number of environmental risks is a subject of continuing debate and reappraisal. The scene changes and sometimes changes rapidly. At the time of the UN Environment Conference held at Stockholm in 1972 the nations of the world perceived their interests to be at variance with each other in the environmental realm (Falk, 1972). To a remarkable degree the perception of *antagonistic interests* has been replaced with a perception of the *common interests* of mankind. This is a crucial change because it is only as the sense of common interest grows that an international order can be created in which environmental risks are managed rationally and for the benefit and survival of all men.

The range of international risks described in the foregoing pages makes clear that international collaboration needs further strengthening and development to cope with existing problems and the new risks that are certain to emerge as the impact of human activities on the resources of the biosphere continues to grow.

We see a number of avenues for development in this direction. In none of them is the path clear, but all need to be sought out and followed if an effective and successful management of environmental risks is to be achieved internationally.

6.4.1 National Risk Management Institutions

Perhaps of greatest importance is the development and strengthening of national risk management institutions. The style and character of the institutions needs to be commensurate with the problems and priorities of the individual nations. The concept of environmental risk management can only make headway on an international scale when national governments are familiar with the ideas and techniques required and use them in their own areas of jurisdiction. For this reason we think there is need for collaboration between nations and action by international organizations to help all nations develop and strengthen risk management capacities.

This process is in fact going on in many ways through many of the specialized agencies of the UN system. National capacity is best developed on national problems and hence an important area for collaboration is in the category of *common national problems* described above. The UN Disaster Relief Office is helping nations to find better ways of coping with natural hazards and disasters. The World Health Organization plays a leading role alongside others in helping to overcome public health risks associated, for example, with inadequate and unsafe supplies of drinking water. Many other examples of this sort could be cited. The important point here is that strengthening of national risk management capacity is an essential preliminary to the development of more coordinated international responses. The International Referral System being developed and operated through UNEP is an important step in this direction.

6.4.2 International Activities

An important priority at the international level is getting the facts straight. Progress is difficult to achieve as long as there are areas of wide scientific disagreement. A common scientific basis of understanding is difficult to achieve *within* nations and environmental disputes are often characterized by expert testimony from scientific witnesses who disagree. At the international level procedures need to be strengthened for establishing as large an area of agreement among scientists as possible. The use of the International Council of Scientific Unions (and its constituent bodies) as one means of drawing on scientific expertise in a relatively unbiased fashion is one direction that can be followed. For any environmental risk problem that entails physical linkages between countries, the kind of trade-off table produced for sulphur (see Table 6.1) by OECD specialists is a very large step towards effective management. Clearly such scientifically agreed estimates can still be challenged, and their existence does not by itself provide any guarantee that progress towards a solution will follow, but the more authoritative the statements of international scientific groups and the wider the area of agreement, the greater the prospects for international agreement on required actions.

While seeking areas of scientific agreement, there is a hierarchy of steps in international collaboration that can be followed at the same time (Eldin, 1973). First there is need to exchange information on national experience. The more that such exchange can be arranged the better. Useful lessons can be learned in the process of exchange and the exchange itself helps to build common understandings of environmental risks and the *alternatives* open to management.

A second step is to harmonize national decisions. Clearly this requires detailed and sometimes lengthy negotiations. There are many examples where national decisions on environmental questions have been harmonized as well as many that have not. One successful mechanism for harmonizing national decisions has been the International Joint Commission established between the United States and Canada for the management of the Great Lakes under the Boundary Waters Treaty of 1909 (Ross, 1972). This Commission is composed of six members, three from each country, who try to act in unison to achieve the best solution in the *common interest* of the two countries. Thus in the words of former Chairman Heeney, we 'act, not as delegates striving for national advantages under instructions from their respective governments, but as members of a single body' (Heeney, 1966). The method of harmonizing national decisions necessarily varies according to the situation and the nations involved, but the need to seek such arrangements is now widely recognized.

A third and more difficult step is to integrate environmental management policy with international trade and development policy. The danger of policies adopted in the name of environmental protection becoming, in effect, non-tariff barriers to international trade is very real and serious. As matters of environmental risk assume larger significance in the affairs of nations, as we believe they must, then there is increasing need for a voice representing the common interests of mankind in a safe environment to be heard in the councils of international economic affairs. To the extent that this voice lacks authority, the environmental future will be more in jeopardy from short-run and sectional economic interests. Serious consideration is therefore needed about ways and means to ensure that environmental risks are considered at the highest levels of international discussion on economic affairs.

There is hidden in the descriptions of the ways in which risk assessments and risk management enters into relations between nations a thinly veiled implication of massive proportions. What happens when a truly global risk of serious and pressing proportions is established? This *could* turn out to be the case with the ozone layer depletion risk. It *could* turn out to be the case with changes in the carbon cycle and the associated risk of climatic change or instability. It *could* turn out to be the case with any one of a number of chemical products, the use of which has become or will become heavy and widespread on a global basis.

These risks are present now, but they are all sufficiently uncertain or apparently far enough into the future that serious action can be delayed. But not forever. From the point of view of scientific risk assessment, and from the perspective offered by the present spectrum of environmental risks, it seems to be only a matter of time before a truly global risk of serious and *pressing* proportions appears. This is not an overdramatization. It is a fact of life.

The international mechanisms that now exist and the institutional procedures currently available seem unlikely to be able to respond effectively. The difficulties will be greater the more that the distribution of the risks and benefits are seen to be unequal among the nations.

Recent experience with efforts at concerted international action may be viewed as a learning process and as a training ground for the more serious and urgent decisions that will be required in the future. The United Nations Conference on the Law of the Sea; the World Plan of Action for the Control of Desertification; and the International Drinking Water Decade, are examples of current progress in international cooperation.

Further development in two directions seems indicated:

(1) Attention must increasingly be given to the formulation of a series of step-by-step actions and the mechanisms to be used to safeguard and assert the common interests of mankind *even* when, or more correctly, *especially* when they conflict with the short-term interests of some nations.

(2) As a prerequisite to such action, the capacity of the international scientific community to provide unbiased reports reflecting as large an area of scientific agreement as possible, needs to be greatly strengthened.

Scientific collaboration is one important strategy in combatting environmental risks. But perhaps equally important is a wider understanding of the part of all those involved in risk assessment of the multifaceted nature of the environmental problems that we face. This report has sought to show how risk assessors can move away from responding to hazards in an *ad hoc* fashion towards actions that are more systematic, more accountable and more anticipatory.

References

Aiken, A.M., Harrison, J.M. and Hare, F.K. (1977) *The Management of Canada's Nuclear Wastes,* Ottawa, Energy, Mines and Resources, Energy Policy Sector, Report EP77-6.

Ambio (1976) Special issue on the 'Sulphur Cycle', *5*(2), Stockholm, Royal Swedish Academy of Sciences.

Anderson, F.R. and Daniels, R.M. (1973) *NEPA in the Courts: A Legal Analysis of the Natural Environmental Policy Act,* Resources for the Future, Baltimore, Johns Hopkins.

Ashby, Lord Eric (1975) *Politics and the Environment,* Ditchley Foundation Lecture, Mimeograph.

Baker, R. (1976) 'The administrative trap', *Ecologist, 6*(7), 247-251.

Beg, M. Arshad Ali (1977) 'Environmental Risk Assessment in Pakistan', *SCOPE/ UNEP International Working Seminar,* Tihany, Hungary, June 8-14, 1977, Background Paper No. 13.

Biological Effects of Ionizing Radiation Report (BEIR) (1973) *Bulletin of the Atomic Scientists,* March, 47-49.

Blodgett, J.E. (1974) 'Pesticides: regulation of an evolving technology', in S.S. Epstein, and R.D. Grundy (Eds.), *Consumer Health and Product Hazards — Cosmetics and Drugs, Pesticides, Food Additives, Volume 2 of The Legislation of Product Safety,* Chapter 4, pp.198-287.

Burton, I. (1977) 'Safe water for all', *Natural Resources Forum, 1,* 95-110, New York, United Nations.

Butler, G.C. (Ed) (1978) *Principles of Ecotoxicology,* SCOPE Report 12, Chichester, Wiley.

Cohen, B.L. (1978) 'Saccharin: the risks and benefits', *Nature, 271,* 9 February.

Critchley, O.H. (1976) 'Risk prediction, safety analysis and quantitative probability methods — a caveat', *J. Brit. Nuc. Energy Soc., 1,* 15 January, 18-20.

Davis, S.M. and Lawrence, P.R. (1977) *Matrix,* Addison-Wesley Series on Organisation Development.

d'Arge, Ralph C. and Allen V. Kneese 'Environmental Quality and International Trade' *International Organization*, Vol 26, No.2, Spring 1972.

Doern, G.B. (1977) *Regulatory Processes and Jurisdictional Issues in the Regulation of Hazardous Products in Canada,* Ottawa, Science Council of Canada, Background Study No.41.

Eckholm, E.P. (1976) *Losing Ground: Environmental Stress and World Food Prospects,* New York, Norton.

Eldin, Gerald (1973) 'The need for intergovernmental cooperation and coordination in environmental policy', in Allen V. Kneese, S.E. Rolfe and J.W. Harned (Eds.), *Managing the Environment: International Economic Cooperation for Pollution Control,* New York, Praeger, pp.199-207.

Ehrlich, P., Ehrlich, A. and Holdren, P. (1977) *Ecoscience: Population, Resources, Environment,* W.H. Freeman and Company.

Evans, W.J. and Hare, F.K. (1979) *Stratospheric Pollution* (to be published).

146

Falk, Richard A. (1972) 'Environmental policy as a world order problem', *Natural Resources J., 12*(2), April, 161-171.

Fife, P.K. (1973) 'Professional liability and the public interest', in *Risk Acceptance and Public Policy,* Proceedings of Session IV, International System Safety Society Symposium, July 1973, Denver, Colorado.

Fimreite, N. (1969) *Mercury Contamination in Canada and its Effects on Wildlife,* unpublished Ph.D. Thesis, University of Western Ontario, Department of Zoology, London, Ontario.

Fjeld, B. (1976) *Furbruk av fossiy brensel: Europa ag utslipp av SO_2 i perioden 1900-1972.* Kjeller, Norway, Norsk Institutt for Luftforskning, Taknisk notat No. 1.

Frank, N.L. and Hussain, S.A. (1971) 'The deadliest cyclone in history', *Bull. Amer. Met. Soc., 52*(6), 438-444.

Friberg, L. (1976) *PAHO/WHO Workshop on Air Quality Criteria and Standards,* Sao Paulo, Brazil, 1976.

Gibson, S.B. (1976) 'The use of quantitative risk criteria in hazard analysis', *J. of Occup. Acc., 1,* 85-94, Elsevier.

Grundlach, E.R. (1977) 'Oil tanker disasters', *Environment, 19*(9), 16-27.

Hare, F.K. (1977) 'Is the climate changing?', *Mazingira, 1,* 19-29.

Hallberg, R.O. (1976) 'The global sulphur cycle', *in* B.H. Svensson and R. Soderlund (Eds), *Nitrogen, Phosphorus and Sulphur — Global Cycles,* SCOPE Report 7, Swedish Natural Science Research Council, Stockholm.

Harper, F.V. and James, F. (1956) *The Law of Torts,* Little, Brown & Co., Boston, p.936.

Heeney, A.D.P. (1966) *Diplomacy with a Difference,* International Nickel Company Inc., Reprint.

Hernberg, S. (1972) 'Biological effects of low lead doses', *Proceedings International Symposium on Environmental Health Aspects of Lead,* October 1972, Amsterdam, pp.617-629.

Holdgate, Martin W. and White, G.F. (1977) *Environmental Issues,* SCOPE Report 10, London, Wiley.

Inhaber, H.M. (1978) *Risk of Energy Production,* Ottawa, Atomic Energy Control Board, AECB Report 1119.

Irukayama, K. (1966) *3rd International Conference on Water Pollution Research,* Washington, D.C., Water Pollution Control Federation.

Izmerov, N.F. (1973) *Control of Air Pollution in U.S.S.R.,* Geneva, WHO, WHO Public Health Papers No. 54.

Jensen, S. and Jernelov, A. (1969) 'Biological methylation of mercury in aquatic organisms', *Nature, 223,* 753.

Johnson, H., Johnson, J.M. and Gour-Tanguay, R. (1977) *Environmental Policies in in Developing Countries,* Beiträge zur Umwelt-gestaltung Heft A27. Berlin, Erich Schmidt Verlag.

Kassas, M. (1975) 'Arid and semi-arid lands: an overview', in UNEP, *Overviews in the Priority Subject Area: Land, Water and Desertification,* UNEP/PROG/2, February, Nairobi, UNEP.

Kates, R.W. (1978) *Risk Assessment of Environmental Hazard,* SCOPE Report 8, Chichester, Wiley.

Keeling, C.D. and Bacastow, R.B. (1977) 'Impact of industrial gases on climate', *USNAS,* 72-95.

Kellogg, W.W. (1977) 'Effects of human activities on global climate', *WMO Bulletin, 26,* Part 1, 229-240; *27,* Part 2, 3-10.

Kuratsune, M. *et al.* (1972) 'Epidemiological study of Yusho', *Environmental Health Perspectives, 1.*

Lawless, E.W. (1977) *Technology and Social Shock,* New Brunswick, N.J., Rutgers.

Leiftinck, P., Sadove, A.R. and Creyke, T.C. (1969) *Water and Power Resources of West Pakistan: A Study in Sector Planning,* Vol. II, Baltimore, Published for World Bank by Johns Hopkins.

Lin, S.C. *et al.* (1976) 'Limitations of fertilizer induced ozone reduction by the long lifetime of the reservoir of fixed nitrogen', *Geophys. Res. Letters, 3,* 157-160.

Lindstedt, G., and Skorfuing, S. (1972) in L. and J. Vostal (Eds.), *Mercury in the Enviroment,* CRC Press.

Löfroth, G. (1970) 'Methylmercury', *Ecol. Res. Comm. Bull. No. 4* (2nd edition), Stockholm, Swedish National Research Council.

Lowrance, W.W. (1976) *Of Acceptable Risk,* Los Altos, Calif., Kaufmann.

Machta, L. (1976) *The Ozone Depletion Problem,* MARC Report No. 1, London, Monitoring and Research Assessment Centre, Chelsea College, University of London.

Martin, B. and Sella, F. (1977) *International Monitoring Activities, International Environmental Monitoring: A Bellagio Conference,* New York, Rockefeller Foundation.

McGinty, L. (1976) 'Whose acceptable risk?', *New Scientist,* 16 September, 582-583.

Mishan, E.J. (1976) *Cost-Benefit Analysis* (new edition), New York, Praeger.

Mootooka, P.S. (Ed.) (1977) *Proceedings of Conference on the Impact of Pesticide Laws,* December 6-10, 1976, Honolulu, Hawaii, East-West Center.

Munn, R.E. (1973) *Global Environmental Monitoring System: GEMS,* SCOPE Report 3.

Munn, R.E. (Ed.) (1977) *Stratospheric Ozone Depletion: An Environmental Impact Assessment,* A Report prepared for UNEP by ICSU-SCOPE through the Monitoring and Assessment Research Centre, Chelsea College, University of London, 15 January (Unpublished).

Munn, R.E., Phillips, M.L. and Sanderson, H.P. (1977) 'Environmental effects of air pollution: implications for air quality criteria, air quality standards and emission standards', *The Science of the Total Environment, 8,* 53-67.

New York Times, March 13, 1977.

Newman, J.L. (Ed.) (1975) *Drought, Famine and Population Movements in Africa,* Syracuse, University Press.

O'Brien, R.D. (1967) *Insecticides — Action and Metabolism,* New York, Academic.

Odum, E.P. (1963) *Ecology,* Holt, Rinehart and Winston.

Ontario (1976) *Mercury Poisoning in Iraq and Japan,* Report of a Government of Ontario Visiting Team, June 22, 1976, Toronto, Ontario Government Publications.

Organization for Economic Cooperation and Development (OECD) (1971) *The Problems of Persistent Chemicals,* Report of Study Group on Unintended Occurrence of Pesticides, Paris, OECD.

Organization for Economic Cooperation and Development (OECD) (1977) *The OECD Programme on Long Range Transport of Air Pollutants; Measurements and Findings,* Paris, OECD.

Ottar, Brynjulf (1977) 'International agreement needed to reduce long-range transport of air pollutants in Europe', *Ambio, VI*(5).

Perry, H. and Landsberg, H.H. (1977) 'Project world energy consumption', *USNAS,* 35-50.

Pittock, A.B. (1974) 'Ozone climatology, trends and the monitoring problem', *Proceedings 1st Conference on Lower and Upper Atmospheres and Possible Anthropogenic Perturbations,* IAMAP/IAPSO First Special Assemblies, Melbourne, Australia, 455-466.

Pochin, E.E. (1974) 'Occupational and other fatality rates', *Comm. Health, 6*(2), 2-13.

Prosser, W.L. (1964) *Handbook of The Law of Torts,* The West Publishing Co. St. Paul.

ReVelle, C. and Revelle, P. (1974) *Source Book on the Environment,* Boston, Houghton Mifflin.

Ross, Charles R. (1972) 'National sovereignty in international environmental decision', *Nat. Res. J., 12*(2), April, 242-254.

Roy, B.B. and Pandey, S. (1970) 'Expansion or contraction of the Great Indian Desert', *Proceedings Indian National Science Academy,* **36B,** No.6, 343.

Scientific Committee on Problems of the Environment (SCOPE) (1977) *Report on an International Working Seminar, June 8-14, 1977, Tihany, Hungary,* Mimeograph.

Sinclair, T.C. (1972) *A Cost-Effectiveness Approach to Industrial Safety.* (Committee on Safety and Health at Work. Research Paper) London. Her Majesty's Stationery Office.

Sinclair, G., Marstrand, P. and Newick, P. (1972) *Innovation and Human Risk,* London, Centre for Study of Industrial Innovations.

Soderlund, R. and Svensson, B.H. (1976) 'The global nitrogen cycle', in B.H. Svensson and R. Soderlund (Eds), *Nitrogen, Phosphorus and Sulphur — Global Cycles,* SCOPE Report 7, Swedish Natural Science Research Council, Stockholm.

Sri Vatsa, Laxmipurom P. (1977) 'Oil refinery near the Taj Mahal', *Science, 197,* 23 September,

Stannard, A.F.B. (1969) 'How you risk your life — a study of comparative risks', *Insurance,* October 25, 29-32.

Starr, C. (1972) 'Benefit-cost studies in sociotechnical systems', in *Perspectives on Benefit-Risk Decision Making,* Washington, D.C.: National Academy of Engineering.

Stone, C.D. (1974) *Should Trees Have Standing?: Towards Legal Rights for Natural Objects,* Toronto, Kaufmann.

Treshow, M. (1976) *The Human Environment,* McGraw-Hill.

UK Health and Safety Commission (1978) *The Hazards of Conventional Sources of Energy,* London, HMSO.

UK Health and Safety Commission Advisory Committee on Major Hazards (1974) *Inquiry into the Flixborough Accident,* September, London, HMSO.

United Nations Environment Program (UNEP) (1978) *Ozone Layer Bulletin, 1,* January, Nairobi, UNEP.

United Nations Environment Program (UNEP) (1977) *World Plan of Action on the Ozone Layer,* UNEP/WG.7/24/Rev.1, 7 March.

UNESCO (1975) *Regional Meeting on Integrated Ecological Research and Training Needs in the Sahelian Region,* UNESCO Man and the Biosphere Report No. 18, Paris, UNESCO.

US Agency for International Development, Office of Science and Technology (1972) *Desert Encroachment on Arable Lands: Significance, Causes and Control,* TA/OST 72-10, August, Washington, D.C.

US Atomic Energy Commission (1974) *Reactor Safety Study,* (Wash 1400) Washington, D.C.

US Environmental Protection Agency (1974) *National Policy on the Control of Carcinogens for the Protection of Human Health,* report prepared by EPA with NCI revision, December, Washington, D.C.

US Environmental Protection Agency (1975) *D.D.T.: A Review of Scientific and Economic Aspects of the Decision to Ban its use as a Pesticide,* Washington D.C. July EPC-540/1-75-022, 300p.

US 29 Federal Register 7728, June 17, 1964, Washington, D.C.

US National Academy of Sciences (USNAS) (1977a) *Energy and Climate,* Studies in Geophysics (Report of Panel on Energy and Climate, R.R. Revelle, Chairman) Washington, D.C.

US National Academy of Sciences (USNAS) (1977b) *Halocarbons: Environmental Effects of Chlorofluoromethane Release* (Report of Committee on Impacts of Stratospheric Change, J.W. Tukey, Chairman.) Washington, D.C.

US National Academy of Sciences (USNAS) (1975) *Petroleum in the Marine Environment,* Washington, D.C., National Academy of Sciences.

US Senate (1960) Hearings before the Anti-trust Sub-committee on the Judiciary (Kefauver Sub-committee), U.S. Senate, 1960, pursuant to S.Res.238 at 12040. Testimony by Dr. Barbara Moulton, former MDA medical officer.

US Surgeon General (1964) Advisory Committee on Smoking and Health, *Report 1964,* Washington, D.C.

Warren, D.V. (1977) 'Accident rates and acceptable risk levels for aeroplanes, and their systems', in *Acceptability of Risks,* UK Council for Science and Society, Report of Working Party, London.

Wassermann, M., Tomatis, L. and Wasserman, Dora (1975) 'Organochlorine compounds in the general population of 1970's and some of their biological effects', *Pure and Applied Chemistry, 42,* 189-208

Weinberg, A.

Whyte, A.V.T. (1977) *Guidelines for Field Studies in Environmental Perception,* MAB Technical Notes 5, Paris, UNESCO.

Wollan, M.J. (1968) 'The process of setting safety standards in the courts, congress and administrative agencies', *Staff Discussion Paper 204,* Washington, D.C., Program and Policy Studies in Science and Technology, George Washington University.

Woodwell, G.M. (1977) 'The carbon dioxide question', *Scientific American, 238,* 34-43.

Woodwell, G.M., Craig, P.P. and Johnson, H.A. (1977) 'DDT in the biosphere — where does it go?', *Science, 174,* 1101-1107.

WHO Expert Committee on Atmospheric Pollutants (1964) *Report,* Geneva, World Health Organization, Technical Report Series No.271.

WHO (1972) *Air Quality Criteria and Guides for Urban Air Pollutants,* Report of a WHO Expert Committee, Geneva, World Health Organization, Technical Report Series No. 506.

WHO (1976) *Statement on Modification of the Ozone Layer due to Man's Activities,* WHO, Geneva, 6p.

International Working Seminar on Environmental Risk Assessment in an International Context: Emerging Needs and a Suggested Procedure
Institute of Biology, Tihany, Hungary. June 8-14, 1977

LIST OF PARTICIPANTS

M. Arshad Ali Beg Principal Scientific Officer, Pakistan Council of Scientific and Industrial Research Laboratories, Off. University Road, Karachi-39, Pakistan

G. Bora Department of Geography and Regional Planning, Karl Marx University of Economics 1095, Budapest, Dimitrov tér 8, Hungary

Ian Burton Institute for Environmental Studies, University of Toronto, Toronto M5S 1A4, Ontario, Canada

Phylis Daly Environmental Protection Agency, 401 M Street S.W. Washington, D.C. U.S.A.20460

F.R. Farmer United Kingdom Atomic Energy Authority, Wigshaw Lane, Culcheth, Warrington WA3 4NE United Kingdom

Y. Fukushima 2-7-24 Taishido, Setagaya-Ku, Tokyo 154, Japan

B.W. Garbrah Environmental Protection Council, Parliament House, Accra, Ghana

E. Hadac Institute of Landscape Ecology, Czechoslovakian Academy of Sciences, Pruhovice v Prahy, Czechoslovakia

G.C.N. Jayasuriya Secretary-General, National Science Council of Sri Lanka, 47/5 Maitland Place, Colombo 7, Sri Lanka

R. Jelinek Czechoslovak Academy of Sciences, Institute of Medicine, Lejekova 61, 120 00 Praha 2, Czechoslovakia

Tore K. Jensen Institute of Transport Economy, Postboks 6110 Etterstad, Oslo 6, Norway

P. Kazasov Institute of Hygiene and Occupational Health Medical Academy, Sofia, Bulv. D. Nestorov 15, Bulgaria

Ashok Khosla United Nations Environment Program, P.O. Box 30552, Nairobi, Kenya

Huxley Knox-Macauley Ministry of Health, Freetown, Sierra Leone

V.F. Krapivin Computer Centre of Academy of Sciences, Laboratory Application of Mathematics. Moscow, Vavilova str. 40. Soviet Union

István Láng Deputy Secretary-General, Hungarian Academy of Sciences, Roosevelt Terrace 9, 1361, Budapest, V. Hungary

T.R. Lee Department of Psychology, University of Surrey, Guildford, Surrey GU2 5XH, United Kingdom

Joanne Linnerooth IIASA-IAEA P.O. Box 590, 1010 Vienna, Austria

Lennart J. Lundqvist Department of Government, University of Uppsala, Skytteanum, Box 514, S-75120, Uppsala, Sweden

R.E. Munn Atmospheric Environment Service, 4095 Dufferin Street, Downsview M3H 5T4, Ontario, Canada

A. Oudiz Commissariat á l'Energie Atomique, Centre l'Etudes Nucléair de Fontenay aux Roses, BP No 6 92260 Fontenay aux Roses, France

Prom Panitchpakdi Secretary-General, National Environment Board, 240 Suriyophai Building, Phpholyothin Road, Bangkok, Thailand

E. Pattantyus Deputy Head of the Secretariate of National Council for Environmental Protection at the Government, 1054, Budapest, Belioannis u 2-4, Hungary

Jan Pinowski Institute of Ecology, Polish Academy of Sciences, 05-150 Lomianki, Poland

V. Puscariu Academy of Sciences of the Rouman Socialist Republic, Bucuresti, Roumania

Jorge Rabinovich Instituto Venezolano de Investigaciones Cientificas (IVIC) Centro de Ecologia, Apartado 1827, Caracas 101, Venezuela

Hussein Razavi Plan and Budget Organization, Teheran, Iran

Ralph Richardson Rockefeller Foundation, 111 West 50th Street, New York, N.Y. USA 10020

Hans Rieger South Asia Institute, University of Heidelberg, Heidelberg, Federal Republic of Germany

G. Rzhanova Representative of the Council for Mutual Economic Assistance (CMEA) Moscow, Kalinin prospect 56, Soviet Union

Andrew Sors Monitoring and Assessment Research Centre (MARC) Chelsea College, The Octagon Building, 459A Fulham Road, London SW10 0QX, United Kingdom

P. Stefanovits University of Agriculture, Department of Geology, Professor. H-2301 Gödöllö, Hungary

A. Szesztay Karl Marx University of Economics, Department for Economy Planning, 1828, Budapest, 5 Pf.489, Hungary

Marta Ventilla Program Officer, Ford Foundation, 320 East 43rd Street, New York, N.Y. USA 10020

G. Vida Head of Dept of Genetics, Eötvös Lóránd University, Budapest, Muzeum krt 2, Hungary

M. Wasserman Head, Department of Occupational Health, The Hebrew University-Hadassah Medical School, P.O. Box 1172, Jerusalem, Israel

Gilbert F. White Natural Hazard Research Institute of Behavioral Science, University of Colorado, Boulder, Colorado, USA. 80309

Anne Whyte Institute for Environmental Studies, University of Toronto, Toronto M5S 1A4, Ontario, Canada

Index

154

156

International Council of Scientific Unions (ICSU)
Scientific Committee on Problems of the Environment (SCOPE)

SCOPE is one of a number of committees established by a non-governmental group of scientific organizations, The International Council of Scientific Unions (ICSU). The membership of ICSU includes representatives from 68 National Academies of Science, 18 International Unions, and 12 other bodies called Scientific Associates. To cover multidisciplinary activities which include the interests of several unions. ICSU has established 10 scientific committees, of which SCOPE, founded in 1969, is one. Currently, representatives of 34 member countries and 15 Unions and Scientific Committees participate in the work of SCOPE, which directs particular attention to the needs of developing countries.

The mandate of SCOPE is to assemble, review, and assess the information available on man-made environmental changes and the effects of these changes on man; to assess and evaluate the methodologies of measurement of environmental parameters; to provide an intelligence service on current research; and by the recruitment of the best available scientific information and constructive thinking to establish itself as a corpus of informed advice for the benefit of centres of fundamental research and of organizations and agencies operationally engaged in studies of the environment.

SCOPE is governed by a General Assembly, which meets every three years. Between such meetings its activities are directed by the Executive Committee.

Frank Fenner
Editor-in-Chief
SCOPE Publications

Executive Secretary: Dr V. Smirnyagin

Secretariat: 51 Bld de Montmorency
75016 PARIS